Basic Knowledge
'O' Grade Mathematics

J. F. Morgan

Robert Gibson & Sons, Glasgow, Ltd.
17 Fitzroy Place, Glasgow, G3 7SF

Contents

READ THIS FIRST

This is your notebook. Keep it by you and refer to it regularly. When you find an example difficult, read the relevant section till you understand it. Don't be satisfied by following any example or explanation line by line, but study it until you understand why that particular step was taken at that time when so many other steps might just as well have been taken. It is not *how* the example is unfolded line by line but *why* that is important. If you follow mathematics by asking "why?" instead of "how?" you will soon find you have more mathematical ability than you have been aware of.

Concentrate your study on one specific problem, or type of problem at a time, till you master it, then move on, because what follows will depend on what has come earlier. Regular, short periods of concentration will pay greater dividends than long, tedious spells every now and then. Keep your work neatly laid out with the working for each step close by, alongside the logical steps you are following. This way you are less likely to lose track of your arguments, and you will save time in checking for errors. If you find you cannot go any further in a particular example, leave a space and start another, then come back for a fresh look, for maybe what you have done since your first attempt will give a clue to the next step in the unfinished problem.

Read your problem carefully to be sure you are aware of all the information that is being given, and know exactly what you are being asked to do. Plan your answer, and write down what you are required to find before you attempt the solution. This way you break the problem down into simple steps, each of which you probably can solve, and with this overall picture of the problem you have a much better chance of seeing where it leads.

When you are studying, make sure that before you start you have all you will require at hand e.g. tables, protractor, compasses etc. so that your concentration will not be broken by having to search them out half way through the argument. Find a comfortable place to work where you are unlikely to be disturbed, and always study there. Soon you will find study will become a habit more than a chore and, with increased ability, your confidence will improve your memory and your prospects.

"Practice" is the keyword—practise what is at the time a difficulty. Revise regularly till there is no longer a burden on your memory but a set of automatic techniques which guarantee success. You will find it a good investment to practise test papers which will familiarise you with what you are preparing to meet. These papers may be obtained from R. Gibson & Sons, Glasgow, Ltd., 17 Fitzroy Place, Glasgow G3 7SF.

Printed in Great Britain by
Robert MacLehose & Co. Ltd
The University Press, Glasgow
© J. F. Morgan 1970
ISBN 0 7169 6919 X

Algebra

Notation

E   the universal set.

Ø   the empty set (or { })

N   the set of natural numbers $\{1, 2, 3, \ldots \ldots\}$

W   the set of whole numbers or counting numbers. $\{0, 1, 2, 3, \ldots \ldots\}$

Z   the set of integers $\{\ldots \ldots -3, -2, -1, 0, 1, 2, 3, \ldots \ldots\}$

Q   the set of rational numbers $\{\frac{p}{q} : p, q, \in Z, q \neq 0\}$

R   the set of real numbers. A real number is one which corresponds to a point on the number line.

If we take R as the universal set then all the above sets are subsets of R (see fig. 1).

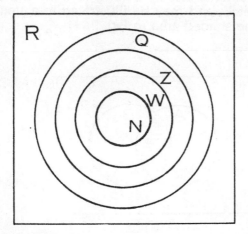

Fig. 1

Sets

A set is a collection of elements and may be denoted by a capital letter, by the enumeration of the elements, or by describing the properties required to be an element of the set. The form $\{ (x, y) : x + y = 3, x, y \in R \}$ is called the set builder notation.

E is the universal set which includes all the elements about which we intend to speak.

Two sets are equal only if they have the same members or elements. The order does not matter unless this is given as a property of the set.

$\emptyset$ or $\{\ \}$ indicates the empty set i.e. the set which has no elements.

$a \in A$ means 'a' is an element of the set A.

$b \notin A$ means 'b' is not an element of the set A.

$\forall$ x means "for all x".

$\exists$ x means "there exists an x".

$A \cap B$ reads A intersection B and denotes the set of elements which belong to *both* the set A *and* the set B, and may be illustrated in a Venn Diagram as the shaded area in Fig. 2.

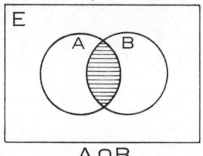

$$A \cap B$$

Fig. 2

A ∪ B reads A union B and denotes the set of elements which belong to the set A *or* the set B *or both,* and is illustrated by the shaded area in fig. 3.

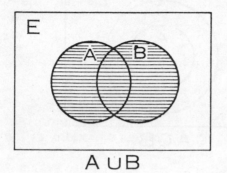

A ∪ B

Fig. 3

A′ is the complement of the set A and is the set of elements which does not belong to the set A. This is illustrated in fig. 4 by the shaded area.

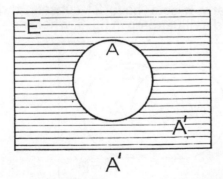

A′

Fig. 4

9

A set may be a subset of another set. $A \subset B$ mean that the set A is a subset of the set B *i.e.* any elemen of the set A is also an element of the set B. This illustrated in Fig. 5.

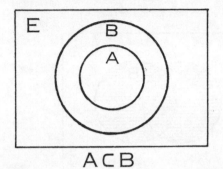

$$A \subset B$$

Fig. 5

$A \subset B$ means A is a proper subset of the set B *i.* there is at least one element of the set B that is nc an element of the set A.

$A \subseteq B$ means that the set A may have the sam elements as the set B *i.e.* $A = B$ is a possibility.

Any set with n elements can be divided into 2 subsets where $\emptyset$ and the set itself are counted a subsets. e.g. $A = \{a, b. c\}$ has subsets.
$\{a, b, c\}; \{a, b\}; \{a, c\}; \{b, c\}; \{a\}; \{b\}; \{c\}; \{ \}.$
This set of subsets is called the power set.

Similarly n sets have 2^n subsets and each can b named as illustrated in fig. 6, where $n = 2$.

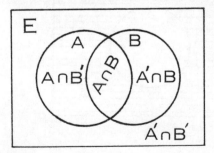

Fig. 6

If n (A) denotes the number of elements in the set A, and n (B) denotes the number of elements in the set B then:—

$$n (A \cup B) = n (A) + n (B) - n (A \cap B)$$

we also have n(C) then:—

$$n(A \cup B \cup C) = n(A) + n(B) + n(C) - n(A \cap B)$$
$$- n(A \cap C) - n(B \cap C) + n(A \cap B \cap C).$$

aws of Sets

Commutative law
$$A \cap B = B \cap A$$
$$A \cup B = B \cup A$$

Associative law
$$(A \cap B) \cap C = A \cap (B \cap C)$$
$$(A \cup B) \cup C = A \cup (B \cup C)$$

Distributive law
$$A \cap (B \cup C) = (A \cap B) \cup (A \cap C)$$
$$A \cup (B \cap C) = (A \cup B) \cap (A \cup C)$$

Idempotent law
$$A \cap A = A$$
$$A \cup A = A$$

Laws with Ø and E
$$A \cap E = A$$
$$A \cup E = E$$
$$A \cap Ø = Ø$$
$$A \cup Ø = A$$

De Morgan's law
$$(A \cap B)' = A' \cup B'$$
$$(A \cup B)' = A' \cap B'$$

Law of absorption
$$A \cap (A \cup B) = A$$
$$A \cup (A \cap B) = A$$

The first identities are said to be the dual of the second identities and vice versa. Notice that the symbol ∩ is changed to ∪ and that E is changed to Ø and vice versa. The complement is unaffected.

An Algebraic Expression

An algebraic expression is a collection of term
separated by $+$ ve and or $-$ ve signs e.g. $4x - y^2$
$4x$ is a term of this expression.
y^2z is a term of this expression.
Each term is composed of factors.
4 is a factor of the term $4x$.
x is a factor of the term $4x$.
$4x$ is a factor of the term $4x$.
We can make '-4' a factor of $4x$ by writing $4x$ in th
form $(-4)(-x)$.
Similarly '$-x$' can be a factor.
We can make '$-4x$' a factor of $4x$ by writing it
the form $-(-4x)$.
Similarly $\pm y$ are factors of the term y^2z.
$\pm y^2$ are factors of the term y^2z.
$\pm y^2z$ are factors of the term y^2z.
$\pm yz$ are factors of the term y^2z.
$\pm z$ are factors of the term y^2z.
x, y, z are called variables of the expression.
constant factor such as '4' in the term $4x$ is calle
the coefficient of the variable x.

Laws

Commutative law (i) $\quad ab = ba$
(ii) $\quad a + b = b + a$
In (i) we may change the *order* of the *factors*.
In (ii) we may change the *order* of the *terms*.
Associative law (i) $\quad (ab)c = a(bc)$
(ii) $\quad (a + b) + c = a + (b + c$
In (i) we may change the *grouping* of the *factors*.
In (ii) we may change the *grouping* of the *terms*.
Distributive law (i) $\quad a(b + c) = ab + ac$
(ii) $\quad ab + ac = a(b + c)$
In (i) we may change a *product of factors into*
sum of terms.
In (ii) we may change a *sum of terms* into a *produ*
of factors.
Note $(a + b)(c + d) = a(c + d) + b(c + d)$
$$= ac + ad + bc + bd$$
Here we have changed a product of factors int
a sum of terms by applying the distributive la
twice.

Identity Elements

Addition of elements on the set R has its own unique identity element, 0 (zero). It may be added to any element of the set R or have any element of the set R added to it so that the sum is equal to the element.

e.g. $a + 0 = a = 0 + a$, $a \in R$

Multiplication of elements on the set R has its own unique identity element, 1. It may be multiplied by any element of the set R or have any element of the set R multiplied by it so that the product is equal to the element.

e.g. $a \times 1 = a = 1 \times a$, $a \in R$

Inverses

The additive inverse is the element on the set R which when added to another element on the set R gives the sum 0 i.e. the identity element of addition.

e.g. $a + (-a) = 0$, $a \in R$. e.g. -8 is the additive inverse of 8.

Each element $a \in R$ has its own unique additive inverse which is also on the set R.

The multiplicative inverse is the element on the set R which when multiplied by another element on the set R gives the product 1 i.e. the identity element of multiplication.

e.g. $a \times \frac{1}{a} = 1$, $a \in R$ e.g. $\frac{1}{8}$ is the multiplicative inverse of 8.

Each element $a \in R$ has its own unique multiplicative inverse which is also on the set R.

13

Mathematical Sentences

"A square is a quadrilateral" is a mathematical sentence. "This is a quadrilateral" is also a mathematical sentence. Now we can decide whether the first sentence is true or false and so this kind of sentence is called a *statement*. The second sentence does not give us enough information to decide whether it is true or false, and this kind of sentence is called an *open sentence*.

$x + 5 = 7$ is an open sentence and we are not able to decide whether it is true or false until we find replacement values for the variable x. An open sentence of this form is called an *equation* and when we find a replacement for x to make such an open sentence into a true statement we say we have *solved* the equation.

The replacement values which *satisfy* the equation i.e. make it a true statement, form the elements of the *solution set* of an equation.

e.g. $x^2 = 9$ has the solution set $\{3, -3,\}$ $x \in Z$ but when $x \in N$ the solution set is $\{3\}$.

However the solution set of $2x = 3$, $x \in W$ is $\{$ $\}$ since there is no whole number which will satisfy the equation.

Linear Equations

Linear equations are those in which the highest power of the variable is unity.

One Variable

Solve $3x + 5 = 17$, $x \in R$

The method of finding the solution set (s.s.) is to employ the additive inverse and the multiplicative inverse as follows:—

$3x + 5 + (-5) = 17 + (-5)$
$\Longleftrightarrow 3x = 12$
$\Longleftrightarrow \frac{1}{3} 3x = \frac{1}{3} 12$
$\Longleftrightarrow x = 4$

The s.s. is $\{4\}$.

When there is only one element in the solution set we say the solution is unique.

Two Variables

When there are two variables and only one equation the s.s. is not unique, as in $y = 2x + 3$, x, y $\in$ R. The solution set of such an equation has an infinite number of elements and they are best represented graphically. The form of the graph of such an equation as $y = mx + c$, x, y, m, c $\in$ R is a straight line, hence the title 'linear equations.'

Method of Drawing the Graph of a Linear Equation

Since a linear equation may be represented graphically by a straight line we need only two points in order to draw the graph of $y = mx + c$. We may calculate the point where the line cuts the y-axis thus, let x = 0.

So $y = m (0) + c$

$<\!\!=\!\!=\!\!=\!\!> y = c$

which gives one of the required points viz. (0, c). We may calculate the point where the line cuts the x-axis thus, let y = 0.

So $0 = mx + c$

$<\!\!=\!\!=\!\!> mx = -c$

$<\!\!=\!\!=\!\!> x = \dfrac{-c}{m}$

This gives the second of the two required points viz. $(\dfrac{-c}{m}, 0)$.

Example: Draw the graph of $y = 2x + 3$. Here $m = 2$, $c = 3$, $\dfrac{-c}{m} = \dfrac{-3}{2}$.

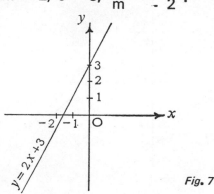

Fig. 7

15

In an equation of the form $y = mx + c$, when c is +ve the line will cut the y-axis c units above the origin. When $c = 0$ the line will go through the origin. When c is —ve the line will cut the y-axis c units below the origin. The value c gives the intercept on the y-axis. Notice that the coefficient of y is unity. All lines with the same value of m are parallel. Fig. 8 illustrates these points.

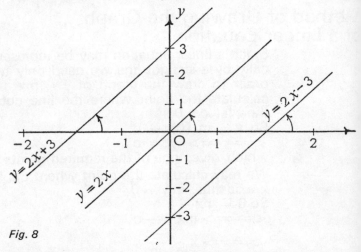

Fig. 8

m is the gradient or slope of the line, and when m is +ve the line will make an acute angle with the +ve direction of the x-axis. If m is —ve the lines will make an obtuse angle with the +ve direction of the x-axis, as in fig. 9.

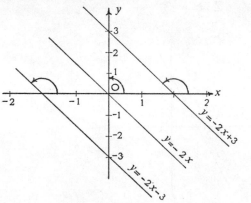

Fig. 9

16

Method of Reading the Equation of a Line from a Graph

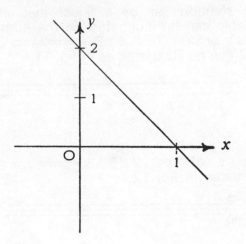

Fig 10

Fig. 10 shows a straight line graph, so its equation is of the form y = mx + c. The line cuts the y-axis 2 units above the origin, so c = 2. The line makes an obtuse angle with the +ve direction of the x-axis, so m is —ve. From this we know the equation is of the form y = mx + 2, m<0. We can now choose a point on the line and substitute the values of x, y into the equation and so find the value of m. The point (1, 0) is on the line and so we can say 0 = m(1) + 2. <====>m = —2. The equation is therefore *y = —2x + 2.*

Linear Inequations

One Variable

The solution set of a linear inequation in one variable can be illustrated on the number line e.g. the s.s. of $3x + 2 < 8$, $x \in R$ <====> $3x < 6$ <=> $x < 2$ can be illustrated as in fig. 11.

Fig. 11

The s.s. of $3x + 2 \leq 8$ <====> $x \leq 2$, $x \in R$ is illustrated in fig. 12.

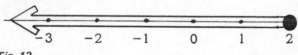

Fig. 12

When the inequation employs the symbols $<$ or $>$ the s.s. does not include the end point(s), and we indicate this by encircling the end point(s), as in fig. 11.

When the inequation employs the symbols $\leq$ or $\geq$ the s.s. includes the end point(s), and we indicate this by encircling and shading the end point(s), as in fig. 12.

Example: $-5 < x \leq 2$, $x \in R$ has s.s. illustrated in fig. 13.

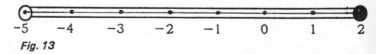

Fig. 13

The s.s. of $-5 < x \leq 2$ is an intersection of 2 sets viz: $\{x : x \leq 2\} \cap \{x : x > -5\}$.

18 Inequations involve the use of symbols $<, \leq, >, >$, and these may often be treated normally ($=$) the same rules applying

Example: $x \leqslant -5$, $x > 2$, $x \in R$ has its s.s. illustrated in fig. 14.

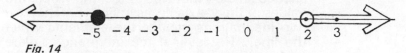

Fig. 14

The solution set of $x \leqslant -5$, $x > 2$, $x \in R$ is a union of 2 sets. viz. $\{x: x \leqslant -5\} \cup \{x: x > 2\}$.

Two Variables

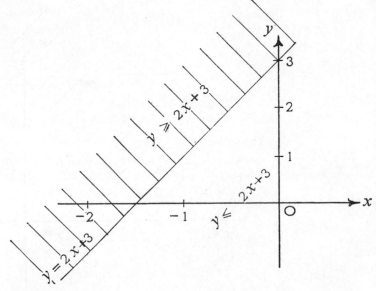

Fig. 15

The elements of the s.s. of the linear equation $y = 2x + 3$ all lie on the line. Sometimes we are more interested in the solution set when $y \leqslant 2x + 3$ or $y \geqslant 2x + 3$ x, y $\in$ R.

When we wish the solution set to the inequation $y \leqslant 2x + 3$ we require the set whose elements all lie in the unshaded area on that side of the line where the lower values of the y-axis lie, as in fig. 15.

twever if both sides of a inaquation are x or ÷ **19**
by a − quantity the signs must change eg
> ÷ <
< ÷ >

When we wish the s.s. to the inequation $y \geqslant 2x + 3$ we require the set whose elements all lie in the shaded area on that side of the line where the greater values of the y-axis lie, as in fig. 15. The s.s. of these two inequations include the elements which lie on the line. However if the inequations were of the form $y > 2x + 3$ and $y < 2x + 3$ we would be excluding the elements which lie on the line, and this would be indicated on the graph by the use of a broken line, as in fig. 16.

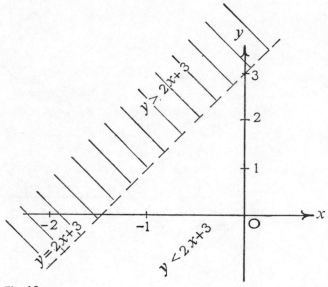

Fig. 16

All the above inequations have assumed $x, y \in R$.
If the inequations required s.s. where $x, y \in Z$ then the diagram would be drawn as in fig. 17, given the inequation $y < 2x + 3$ $x, y \in Z$

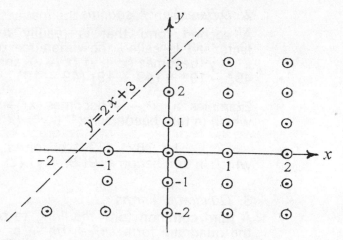

Fig. 17

Note we encircle all the points where x, y co-ordinates are elements of Z, but do not include any point on the line y = 2x + 3.

Factorisation

1. *Common Factor*
Factorisation is the use of the distributive law in reverse. When we factorise completely we express a sum (or difference) of terms as a product of factors e.g. ab + ac is changed into a(b + c). This particular example illustrates the 'taking out of a common factor'. The factor 'a' is common to both terms of ab + ac. When attempting to factorise, this common factor should be dealt with first e.g. 3x + 3y = 3(x + y).

2. *Difference of squares*

A second form that is readily recognised and factorised is called the difference of squares e.g. $x^2 - y^2$ becomes $(x + y)(x - y)$, and $49^2 - 19^2 = (49 + 19)(49 - 19)$.

Examples: a) $x^4 - y^4$ becomes $(x^2 + y^2)(x^2 - y^2)$ which in turn becomes $(x^2 + y^2)(x + y)(x - y)$.

b) $36x^2 - 9y^2$ becomes $9(4x^2 - y^2)$ which in turn becomes $9(2x + y)(2x - y)$.

3. *Quadratic Forms*

A third common form requiring to be factorised is the quadratic form $ax^2 + bx + c$. We shall deal with the case where $a = 1$ *i.e.* $x^2 + bx + c$ which has factors of the form $(x + p)(x + q)$ e.g. $x^2 + 5x + 6 = (x + 3)(x + 2)$
By use of the distributive law twice over
$(x + p)(x + q)$
$\Longleftrightarrow x(x + q) + p(x + q)$
$\Longleftrightarrow x^2 + qx + px + pq$
$\Longleftrightarrow x^2 + (p + q)x + pq$

Notice that the coefficient of x is the sum of the constant terms of the factors *i.e.* $b = (p + q)$.

Notice that the constant term is the product of the constant terms of the factors *i.e.* $c = pq$.

Examples: a) $x^2 + 5x + 6 \Longleftrightarrow (x + 3)(x + 2)$.
Here $(p + q) = 5$, $pq = 6 \Longleftrightarrow$
$p = 3$, $q = 2$.

b) $x^2 - 5x + 6 \Longleftrightarrow (x - 3)(x - 2)$.
Here $(p + q) = -5$, $pq = 6 \Longleftrightarrow p = -3$,
$q = -2$.

Notice that when the constant term is +ve then both factors have the same sign. This sign will be the same as that of the coefficient of x.

When the constant term is —ve then both factors will have different signs, as in the following examples.

a) $x^2 - x - 6 \Longleftrightarrow (x-3)(x+2)$.
Here $(p + q) = -1, pq = -6 \Longleftrightarrow p = -3$, $q = 2$.
b) $x^2 + x - 6 \Longleftrightarrow (x+3)(x-2)$.
Here $(p + q) = 1, pq = -6 \Longleftrightarrow p = 3$, $q = -2$.

Notice that the numerically greater constant in the factors has the same sign as the coefficient of x.

Quadratic Equations

A quadratic equation is of the form $ax^2 + bx + c = 0$. Such a form is often difficult to factorise and thus find the solution set, but there are other methods to help us find the solution set e.g. we may utilise the formula—

$$x = \frac{-b \pm \sqrt{b^2 - 4(a)(c)}}{2a}$$

Example: Solve $2x^2 + 3x + 1 = 0$, $x \in R$
Here $a = +2, b = +3, c = +1$ giving

$$x = \frac{-3 \pm \sqrt{3^2 - 4(2)(1)}}{2.2}$$

$$\Longleftrightarrow x = \frac{-3 \pm \sqrt{1}}{4}$$

$$\Longleftrightarrow x = \frac{-3 + 1}{4} \text{ or } x = \frac{-3 - 1}{4}$$

$$\Longleftrightarrow x = -\tfrac{1}{2} \text{ or } x = -1 \text{ giving s.s.} = \{-\tfrac{1}{2}, -1\}$$

Before employing this formula the coefficient of x^2 must be +ve.
If it is not, then we make it +ve by writing down an equivalent equation.

Example:
Solve $-2x^2 - 3x - 1 = 0$, $x \in R$.
$$-2x^2 - 3x - 1 = 0$$
$$\Longleftrightarrow -1(-2x^2 - 3x - 1) = -1.0$$
$$\Longleftrightarrow 2x^2 + 3x + 1 = 0$$

If the factors can be found for $ax^2 + bx + c$ by inspection then it is quicker to find the solution set as follows.

23

Example : $2x^2 + 3x + 1 = 0$

$\qquad$ <===> $(2x + 1)(x + 1) = 0$

$\qquad$ <===> either $2x + 1 = 0$ or

$\qquad$ $x + 1 = 0$ or both.

$\qquad$ <===> $x = -\frac{1}{2}$ or $x = -1$ or both

Thus the s.s. $= \{-\frac{1}{2}, -1\}$.

Another method of solving a quadratic equation is by completing the square. Recall the form $(x+p)(x+q)$ and consider the case where $p = q$, which would give $(x + p)(x + p)$ or $(x + p)^2$. This latter form is known as a perfect square and on being expressed as a sum of terms gives the quadratic form $x^2 + 2px + p^2$. Quadratic expressions are not always of this form but it is possible to change part of them into a perfect square.

Example : Find s.s. for $2x^2 + 3x + 1 = 0$, $x \in R$.
First make coefficient of x^2 unity.

$\frac{1}{2}(2x^2 + 3x + 1) = \frac{1}{2}. 0$

$\qquad$ <===> $x^2 + \frac{3}{2}x + \frac{1}{2} = 0$

$\qquad$ <===> $x^2 + \frac{3}{2}x = -\frac{1}{2}$

Now compare the left hand side of this equation with the form $x^2 + 2px + p^2$. In our equation

$2p = \frac{3}{2}$ <===> $p = \frac{3}{4}$ ===> $p^2 = \frac{9}{16}$

We may now write our equation as
$(x^2 + \frac{3}{2}x + \frac{9}{16}) = -\frac{1}{2} + \frac{9}{16}$
i.e. we make the left hand side of our equation into a perfect square and keep the balance by adding $\frac{9}{16}$ to the right hand side.

Now $x^2 + \frac{3}{2}x + \frac{9}{16} = \frac{1}{16}$

$\qquad$ <===> $(x + \frac{3}{4})^2 = \frac{1}{16}$

$\qquad$ <===> $x + \frac{3}{4} = \pm\frac{1}{4}$

$\qquad$ <===> $x = -\frac{3}{4} \pm \frac{1}{4}$

$\qquad$ <===> $x = -\frac{3}{4} + \frac{1}{4}$ or $x = -\frac{3}{4} - \frac{1}{4}$

$\qquad$ <===> $x = -\frac{1}{2}$ or $x = -1$

Giving the s.s. $\{-\frac{1}{2}, -1\}$, as before.

Mappings

Mappings arise when there is a relationship between two sets, such that each element in one set is related to only one element in another, e.g.

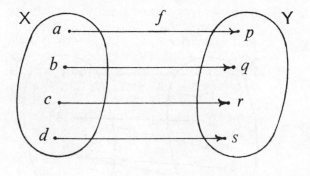

The set X is called the domain. The set Y is called the range. f is the relationship. In this relation we say the set X is mapped *onto* the set Y, because all the elements in the range are images of elements in the domain. In the following mapping the set P is mapped *into* the set Q, since there is at least one element in Q which is not an image of an element in P.

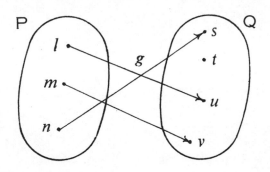

Notice that a mapping must have only one image point for each element in the domain i.e. only one arrow leaves each element in the domain. However, an element in the range may be the image of more than one element in the domain, as in the following example where the relation is "has the square".

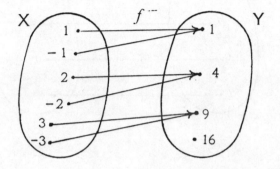

Quadratic Functions

A quadratic function is a mapping of the form $f: x \longrightarrow ax^2 + bx + c$. The set of values of the variable x is called the *domain*. The set of values onto which the elements of the domain are mapped is called the *range*. Each element in the domain is mapped onto only one element in the range. The relationship between the domain and the range may be represented by a graph, e.g. under the mapping $f: x \longrightarrow x^2:-$

f: $0 \longrightarrow 0$

f: $1 \longrightarrow 1$ f: $-1 \longrightarrow 1$

f: $2 \longrightarrow 4$ f: $-2 \longrightarrow 4$

f: $3 \longrightarrow 9$ f: $-3 \longrightarrow 9$

        etc.   etc.

If $x \in Z$ then the domain is the set Z and the range would be the set W.

If $x \in R$ then the range would be the set R^+ i.e. the set of +ve real numbers.

Fig. 18 illustrates a graph of a quadratic function and shows a $\in$ domain mapped onto b $\in$ range.

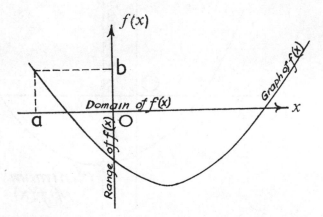

Fig. 18

Quadratic Inequations

The solution set of a quadratic inequation is often readily found from a sketch of the graph of the function. A quadratic inequation is of the form:—

$f(x) < 0$ or $f(x) > 0$
$f(x) \leqslant 0$ or $f(x) \geqslant 0$

This is a shorter way of writing the following:—

$ax^2 + bx + c < 0$ or $ax^2 + bx + c > 0$
$ax^2 + bx + c \leqslant 0$ or $ax^2 + bx + c \geqslant 0$
Since $f(x) = ax^2 + bx + c$

When the coefficient of x^2 is +ve the graph will show a minimum turning point as in fig. 19.

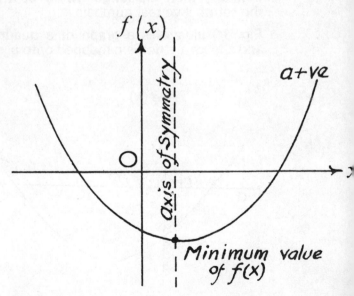

Fig. 19

When the coefficient of x^2 is —ve the graph will show a maximum turning point as in fig. 20.

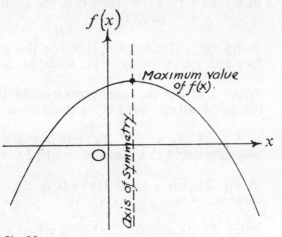

Maximum value of f(x).

Axis of Symmetry

Fig. 20

Notice that the axis of symmetry goes through the maximum or minimum value of $f(x)$.

(i)   (ii)

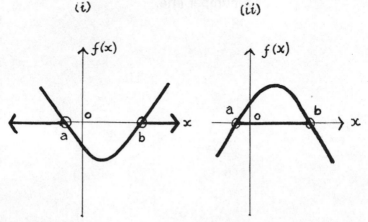

Fig. 21

When $f(x) > 0$ we are interested in the part(s) of the graph which lies above the x-axis.

In fig. 21 (i) s.s. of $f(x) > 0$ is the union of two sets viz. $\{x : x < a\} \cup \{x : x > b\}$.

In fig. 21 (ii) s.s. of $f(x) > 0$ is the intersection of two sets viz. $\{x : x > a\} \cap \}x : x < b\}$ *i.e.* $a < x < b$.

When $f(x) < 0$ we are interested in the part(s) of the graph which lies below the x-axis.

in fig. 21 (i) s.s. of $f(x) < 0$ is the intersection of two sets viz: $\{x : x > a\} \cap \{x : x < b\}$ *i.e.* $a < x < b$.

In fig. 21 (ii) s.s. of $f(x) < 0$ is the union of two sets viz: $\{x : x < a\} \cup \{x : x > b\}$.

In fig. 21 (i) and (ii) $f(x) = 0$ when $x = a$ or $x = b$ *i.e.* $f(x) = 0$ where the graph cuts the x-axis.

Notice that when the part of the graph in which we are interested is in one continuous part the s.s. is an intersection of two sets, but when the part is divided into two separate bits we have a s.s. which is a union of two sets. In either case the solution set may be read off by treating the x-axis as a number line.

Method of Sketching the Graph
of a Quadratic Function

In order to find the solution set of a quadratic inequation we need find only three points on the graph viz. where $f(x) = 0$ and the maximum or minimum values of the function. These can be fixed by completing the square as shown earlier.

Example : Sketch the graph of $f(x) = 2x^2 + 3x + 1$.
$f(x) = 2x^2 + 3x + 1$
$\Longleftrightarrow f(x) = 2(x^2 + \frac{3}{2}x + \frac{3}{4}^2) + 1 - 2. (\frac{3}{4})^2$
$\Longleftrightarrow f(x) = 2(x + \frac{3}{4})^2 - \frac{1}{8}$

Since $2(x + \frac{3}{4})^2 \geqslant 0$, $f(x)$ will have a minimum value when $2(x + \frac{3}{4})^2 = 0$ *ie* when $x = -\frac{3}{4}$. This line $x = -\frac{3}{4}$ is the axis of symmetry of the function. The minimum value of $f(x)$ in our example is then $-\frac{1}{8}$.

$f(x) = 0$ when $2(x + \frac{3}{4})^2 - \frac{1}{8} = 0$
$\Longleftrightarrow 2(x + \frac{3}{4})^2 = \frac{1}{8}$
$\Longleftrightarrow (x + \frac{3}{4})^2 = \frac{1}{16}$
$\Longleftrightarrow x + \frac{3}{4} = \pm\frac{1}{4}$
$\Longleftrightarrow x = -\frac{1}{2}$ or $x = -1$

The sketch of this graph is shown in fig. 22.

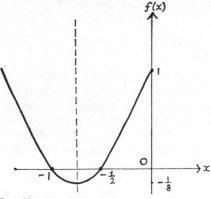

Fig. 22

From fig. 22
$2x^2 + 3x + 1 \geqslant 0$ has s.s. $\{x : x \geqslant -\frac{1}{2}\} \cup \{x : x \leqslant -1\}$.
From fig. 22
$2x^2 + 3x + 1 \leqslant 0$ has s.s. $\{x : x \leqslant -\frac{1}{2}\} \cap \{x : x \geqslant -1\}$
i.e. $-\frac{1}{2} \geqslant x \geqslant -1$.

Implication

If we refer to a figure in a book and say, "If it is a parallelogram then it is a quadrilateral", we are making an implication. Let p stand for "it is a parallelogram", and let q stand for "it is a quadrilateral". Now our implication is "if p then q" or in symbols, p $===>$ q (read, "p implies q"). What we are saying is, that the set of parallelograms forms a subset of the set of quadrilaterals—which is true. If P is the set of parallelograms, and Q the set of quadrilaterals, we may illustrate the implication in a Venn diagram as in fig. 23.

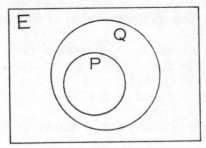

Fig. 23

If the figure to which we referred is a parallelogram, then it must be true that it is a quadrilateral. If the figure to which we referred is not a parallelogram, then it might not be a quadrilateral (it might be a triangle), but no matter what shape it is, our implication will still be true, for we did not say "the figure *is* a parallelogram", but "*if it is* a parallelogram *then* it is a quadrilateral". Although the implication p$===>$q is true its *converse* q$===>$p is not, as may be seen from fig. 23. Some quadrilaterals are not parallelograms. This leads us to the general statement that p$===>$q is true only when the solution set of p (i.e. P) is a subset of the solution set of q (i.e. Q). We should bear in mind that p, q are open sentences.

Example: $x^2 = 25 \implies x = 5$ is a false implication because the solution set of $x^2 = 25$ is $\{5, -5\}$, but the solution set of $x = 5$ is $\{5\}$, so the s.s. of $x^2 = 25$ is not a subset of s.s. of $x = 5$. This example illustrates another way to consider such implications. If we can find at least one example which satisfies the first open sentence, but not the second, then we have proved the implication false. Such an example is called a *counter-example*. The counter-example in the above case is $x = -5$, which satisfies $x^2 = 25$ but not $x = 5$.

Variation

(i) Direct: $x \propto y$, read "x varies directly as y."

(ii) Inverse: $x \propto \dfrac{1}{y}$ read "x varies inversely as y."

(iii) Joint: $x \propto yz$, read "x varies jointly as y and z."

or $x \propto \dfrac{y}{z}$, read "x varies directly as y and inversely as z."

The method for all forms is the same. Once it has been established which kind of variation is involved (which is usually obvious from the context), we can write:

(i) $x \propto y \iff x = ky$ where k is a constant

(ii) $x \propto \dfrac{1}{y} \iff x = \dfrac{k}{y}$

(iii) $x \propto yz \iff x = kyz$

or $x \propto \dfrac{y}{z} \iff x = \dfrac{ky}{z}$

By substituting the given values of x, y, z we can evaluate k. This value of k can then be used throughout the rest of the example.

Surds

A surd e.g. $\sqrt[n]{a}$ is an element of R which cannot be expressed in the form $\frac{p}{q}$ where p, q $\in$ Z, but where "a" can be expressed in this form.   In other words, a surd is the root of a rational number which cannot be expressed as a rational number. A rational number is of the form $\frac{p}{q}$ where p, q $\in$ Z, q $\neq$ o.

Laws of Surds

$$\sqrt{a} + \sqrt{a} = 2\sqrt{a}; \quad \sqrt{a} \times \sqrt{a} = a$$
$$\sqrt{a} \times \sqrt{b} = \sqrt{ab}; \quad m\sqrt{a} \times n\sqrt{b} = mn\sqrt{ab}$$
$$3\sqrt{a} - 2\sqrt{a} = \sqrt{a}$$
$$\frac{\sqrt{a}}{\sqrt{b}} = \sqrt{\frac{a}{b}}; \quad \frac{m\sqrt{a}}{n\sqrt{b}} = \frac{m}{n}\sqrt{\frac{a}{b}}$$

Sometimes it is helpful to *rationalise* a surd when it is in the denominator of a fractional form. This is accomplished by multiplying the numerator and the denominator by the surd which is in the denominator.

e.g. $\frac{4}{\sqrt{2}} <\!\!=\!\!=\!\!> \frac{4\sqrt{2}}{\sqrt{2}\sqrt{2}} <\!\!=\!\!=\!\!> \frac{4\sqrt{2}}{2} = 2\sqrt{2}.$

If we do not rationalise the denominator, then to evaluate $\frac{4}{\sqrt{2}}$ involves dividing 4 by 1·41, and is more complicated than evaluating $2\sqrt{2}$, which is simply the product of 2 and 1·41 i.e. 2·82. Work can often be saved by simplifying a surd. e.g. $\sqrt{50} <\!\!=\!\!=\!\!> \sqrt{25 \times 2} <\!\!=\!\!=\!\!> 5\sqrt{2}$
$= 5 \times 1·41 = 7·05.$

The method of simplifying a surd is to express what is under the square root sign as factors which are perfect squares e.g. 4, 9, 16, 25 etc. If the cube root is involved we look for factors which are perfect cubes e.g. 8, 27, 64 etc.

Simultaneous Equations (2 variables)

When the equations are linear then we need only two of them in order to find a unique solution, if any exists. The reason for this is that two straight lines intersect in only one point (unless they are parallel) provided they are on the same plane. If the lines are parallel then the solution set is Ø. This latter condition is easily recognised since the gradients would be equal.

Example:

$$\left.\begin{matrix} 2x + y = 6 \\ 2x + y = 12 \end{matrix}\right\}$$ is a pair of equations where s.s. = Ø

since both lines have the same gradient and so would be parallel, and would not intersect. If the gradients of the lines are not the same, then the lines would not be parallel, and we could find the point of intersection by drawing a graph, e.g. solve graphically the system of equations $3x - 4y = 2$ and $2x + 3y = 7$.

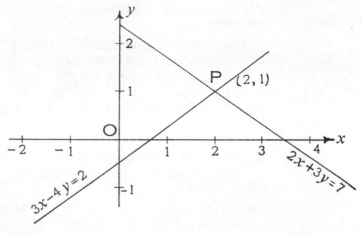

Fig. 24

In fig. 24 the point P (2,1) is the only point common to both lines, and gives the required values of x, y to satisfy the two equations *i.e.* x = 2 and y = 1.

Often the values of x and y are not whole numbers and may not be suited to this method where the degree of accuracy is limited, but there is another method by which such a difficulty may be overcome.

Second method of solving simultaneous equations.

Example: Solve the system of equations

$$2x + 3y = 7$$
$$3x - 4y = 2$$

Method: Find the L.C.M. of the coefficients of the variable we wish to eliminate.

To ELIMINATE x	To ELIMINATE y
L.C.M. of 2 and 3 is 6	L.C.M. of 3 and 4 is 12

Multiply both equations by the appropriate factors of the L.C.M., to give the same numerical value of the coefficient of the variable we wish to eliminate.

$3(2x + 3y) = 3 \times 7$	$4(2x + 3y) = 4 \times 7$
$2(3x - 4y) = 2 \times 2$	$3(3x - 4y) = 3 \times 2$

We *add* the two equations if the coefficient have *different signs*.

We *subtract* the one equation from the other if the coefficients have the *same sign*.

$3(2x+3y)-2(3x-4y)$	$4(2x+3y)+3(3x-4y)$
$= 21 - 4$	$= 28 + 6$
$6x + 9y - 6x + 8y = 17$	$8x + 12y + 9x - 12y = 34$
$17y = 17$	$17x = 34$
$y = 1$	$x = 2$

By substituting the value found back into one of the original equations, we can find the value of the other variable.

$2x + 3(1) = 7$	$3(2) - 4y = 2$
$2x = 4$	$4y = 4$
$x = 2$	$y = 1$

Linear Programming

The efficient use of men and materials can increase production and profits by reducing loss and waste. The ideal is to maximise the former pair and minimise the latter. This ideal is seldom achieved because there are limitations in resources, restrictions to be met, and rules and regulations to be obeyed.

Linear programming is a powerful means of solving problems involving many variables and restrictions which can be brought together to form a set of linear inequations. Computers can then be used to solve such problems, but simpler problems involving only two variables may be solved by means of graphing the inequations on a Cartesian diagram.

The idea behind linear programming is to translate the variables and restrictions into a set of inequations, thus forming what is known as a *mathematical model* of the problem. Having done this the next step is to graph the inequations on a diagram, and thus find the intersection of the solution sets of these inequations.

This area of intersection is known as *the area of feasible solutions* i.e. any point in the area would satisfy all the inequations (given $x, y \in R$). Now the *maximum* or *minimum* solution has to be found, and this is done by considering the expression which has to be maximised or minimised (called the *objective form*) as a family of parallel lines, some of which will intersect with the area of feasible solutions. If the expression is to be maximised, we choose the line which makes the greatest intercept with the y-axis, and if the minimum is sought, we choose the one which makes the least intercept with the y-axis.

When $x, y \in R$ the maximum or minimum will lie on a vertex of the area of feasible solutions, but if $x, y \in W$ then the solution will be the co-ordinates of the point nearest to the vertex where $x, y \in W$. It is possible that the solution set co-incides with a line on the area of feasible solutions and not just a point.

Method

Bear in mind that there can be only two variables x, y if you are to use a Cartesian diagram, so let x be the number of one kind of item, and let y be the number of the other. All the left hand sides of the inequations will be of the form $ax + by$, and this also will be the shape of the objective form to be maximised or minimised. Next we find the limits or restrictions on time, materials, quantity etc. and these amounts will be the right hand side of the inequations. The signs $<$, $>$, $\leqslant$, $\geqslant$ will be obvious from the wording. The coefficients of x and y will come from information given about the *same* materials in any inequation e.g. if ingredients A, B, C are given, then the first inequation will have coefficients of x, y which are both amounts of A. The second will have both coefficients which are amounts of B, and the third amounts of C. If the cost of manufacture is to be minimised or maximised then the coefficients of x, y in the objective form will be the cost of making that particular article. Similarly if the time is to be minimised the coefficient of x will be the time taken to do that operation, and the coefficient of y will be the time to do the other operation. A table of the following form is often helpful:—

	Items		
	x	y	limits
Ingredients { A			
B			
C			

Example:
A forklift truck moves platforms of boxes which are of two kinds. The larger is 2 ft. wide and weighs 8 cwt., and the smaller is 3 ft. wide and weighs 6 cwt. Both are 6 ft. long and 6 ft. high, and the lengths must lie the one way on the platforms. No boxes can be stacked on top of another. If the truck can lift up to 32 cwt., and the platform is 12 ft. wide, what is the maximum number of boxes that can be taken on each lift?

	Large X	Small Y	Restrictions
Width	2x	3y	⩽12
Weight	8x	6y	<32

$$x \geqslant 0 \qquad y \geqslant 0 \qquad X, y \in w$$
$$x + y \text{ to be maximised.}$$

The inequations are
$$2x + 3y \leqslant 12$$
$$8x + 6y < 32$$

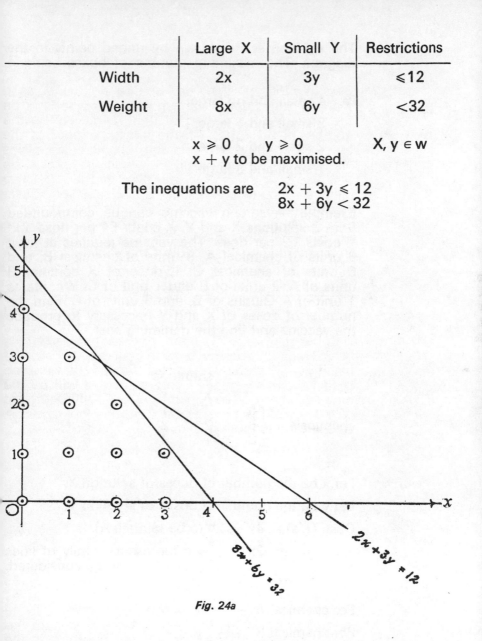

Fig. 24a

The solution set is shown by ringed points in the diagram. The maximum number of boxes being 4

i.e. 4 small and no large.

3 small and 1 large.

2 small and 2 large.

1 small and 3 large.

Example: A certain vaccine can be compounded from 2 solutions X and Y. X costs £4 per dose and Y costs £2 per dose. The vaccine requires at least 6 units of chemical A, 8 units of chemical B, and 6 units of chemical C. 1 dose of X contains 3 units of A, 2 units of B and 1 unit of C. Y contains 1 unit of A, 2 units of B, and 3 units of C. Find the number of doses of X and Y necessary to produce the vaccine and find the minimum cost.

		solutions			
		X	Y		limits
	A	3	1	$\geqslant$	6
chemicals	B	2	2	$\geqslant$	8
	C	1	3	$\geqslant$	6

Let x be the number of doses of solution X.

Let y be the number of doses of solution Y.

Cost (£'s): $4x + 2y$ to be minimised.

i.e. $2x + y = c$ represents family of lines to be considered.

For chemical A $3x + y \geqslant 6$

For chemical B $x + y \geqslant 4$

For chemical C $x + 3y \geqslant 6$

Since the amounts cannot be —ve, x, y $\geqslant$ 0.

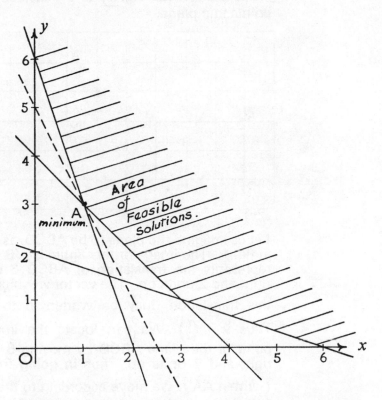

Fig. 25

From fig. 25 the minimum is represented by the point A(1, 3) i.e. 1 dose of X is required and 3 doses of Y, giving a minimum cost of £(4 × 1 + 2 × 3) = £10.

Geometry

Vectors

A vector (written e.g. $\underline{v}$) is a translation of all th points in a plane.

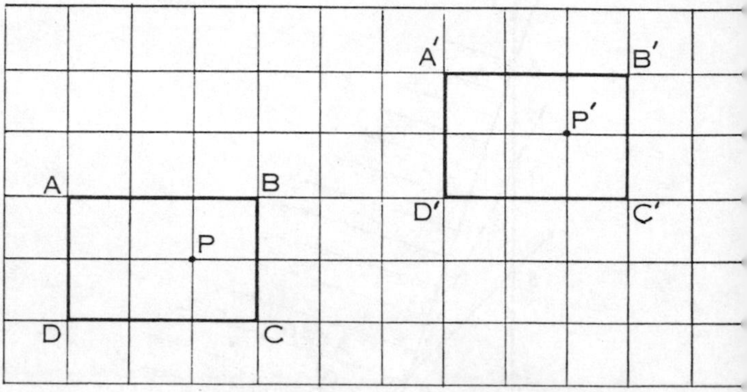

Fig. 1

Let us imagine the plane to be ABCD, as represente in fig. 1. The image of this figure is A'B'C'D', whic represents the translation of ABCD 6 units to th right and 2 units up. The vector $\underline{v}$, which represent this translation, may be written in the form $\binom{6}{2}$

Thus $\underline{v} = \binom{6}{2}$. We can locate the image of an point of the plane ABCD by moving 6 units to th right and 2 units up. Thus in going from A to A (written $\overrightarrow{AA'}$), we move according to the instructio $\binom{6}{2}$.

Thus $\overrightarrow{AA'} = \binom{6}{2}$.

Similarly $\overrightarrow{PP'} = \binom{6}{2}$.

$\overrightarrow{AA'}$, $\overrightarrow{PP'}$ etc. are called directed line segments, an the vector $\underline{v}$ is represented by such a directed lin segment. $\overline{v}$ is in fact the set of all the directed lin segments going from every point in the plan ABCD to its image point in A'B'C'D'.

Thus $\underline{v} = \{\overrightarrow{AA'}, \overrightarrow{BB'}, \ldots, \overrightarrow{PP'}, \ldots \overrightarrow{CC'}, \overrightarrow{DD'}.\}$

Any element in this set can be taken to represent the vector **v**.

If each point in the plane ABCD is joined to its image then the directed line segments look as in fig. 2.

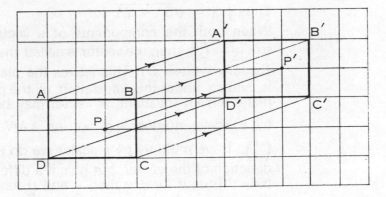

Fig. 2

All the directed line segments have the same length and so are said to be equal in *magnitude*.

All the directed line segments are parallel and so are said to have the same *direction*.

All the directed line segments are going from ABCD to A'B'C'D' and are said to have the same *sense*.

A vector then has magnitude, direction and sense and each of its representatives will have the same magnitude, direction and sense.

Thus $\vec{AA'} = \vec{BB'} = \vec{CC'}$ etc.

In general $\underline{v} = \underline{u}$ only if they have the same magnitude, direction and sense.

In $\underline{v} = \binom{6}{2}$ the '6' and the '2' are called the *components* of $\underline{v}$.

The length or magnitude of a vector (written $|\underline{v}|$ or $|\vec{AA'}|$) is the square root of the sum of the squares of the components.

Thus $|\underline{v}| = \sqrt{6^2 + 2^2}$.

When both the components of a vector are zero e.g. $\underline{w} = \binom{0}{0}$, then the vector is called the zero vector (usually written O). This leaves the plane where it is, and so, under this translation, all the points in the plane remain invariant. A vector may be multiplied by a scalar number e.g. $k\underline{v}$ or $k\vec{AA'}$ or $k\binom{6}{2} = \binom{k6}{k2}$. If we multiply by a scalar we do not alter the direction of the vector, but give it a different magnitude. Thus if $k\underline{v} = \underline{u}$ then $\underline{v}$ and $\underline{u}$ have the same direction i.e. they are parallel.

$\vec{AA'} = \binom{6}{2}$ but $\vec{A'A} = \binom{-6}{-2}$ i.e. 6 units to the left and 2 units down.

i.e. $\vec{A'A} = \binom{-6}{-2} = \binom{-1.6}{-1.2} = -1\binom{6}{2} = -\vec{AA'}$.

Thus $\vec{A'A} = -\vec{AA'}$.

$\vec{AA'}$ is equal in magnitude to $\vec{A'A}$.

$\vec{AA'}$ has the same direction as $\vec{A'A}$ i.e. they are parallel.

$\vec{AA'}$ has the *opposite sense* to $\vec{A'A}$.

In general $k\underline{v} = \underline{u}$ tells us that the vectors $\underline{v}$ and $\underline{u}$ are parallel i.e. they have the same direction. If k is +ve the vectors will have the same sense. If k is —ve the vectors will be opposite in sense. If $k = \pm 1$ the vectors will have the same magnitude.

Addition of Vectors

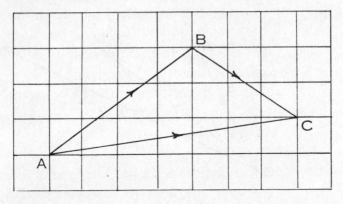

Fig. 3

In fig. 3 $\vec{AB} = \begin{pmatrix} 4 \\ 3 \end{pmatrix}$, $\vec{BC} = \begin{pmatrix} 3 \\ -2 \end{pmatrix}$, $\vec{AC} = \begin{pmatrix} 7 \\ 1 \end{pmatrix}$.

In going from A to B and then to C we would have the same result as going from A directly to C. Since this is the case, we may say $\vec{AB} + \vec{BC} = \vec{AC}$, or

$$\begin{pmatrix} 4 \\ 3 \end{pmatrix} + \begin{pmatrix} 3 \\ -2 \end{pmatrix} = \begin{pmatrix} 4+3 \\ 3+(-2) \end{pmatrix} = \begin{pmatrix} 7 \\ 1 \end{pmatrix}.$$

Note the shape $\vec{AB} + \vec{BC} = \vec{AC}$.
In general $\vec{PQ} + \vec{QR} = \vec{PR}$.

Subtraction of Vectors

Subtraction is performed as follows:
$v - u$ becomes $v + (-u)$

e.g. $\vec{AC} - \vec{AB} = \vec{AC} + (-\vec{AB}) = \vec{AC} + \vec{BA}$

i.e. $\begin{pmatrix} 7 \\ 1 \end{pmatrix} - \begin{pmatrix} 4 \\ 3 \end{pmatrix} = \begin{pmatrix} 7 \\ 1 \end{pmatrix} + \begin{pmatrix} -4 \\ -3 \end{pmatrix} = \begin{pmatrix} 7+(-4) \\ 1+(-3) \end{pmatrix} = \begin{pmatrix} 3 \\ -2 \end{pmatrix} = \vec{BC}$

From fig. 3 Then $\vec{AC} - \vec{AB} = \vec{BC}$.
Again notice the shape and the order of $\vec{AC} - \vec{AB} = \vec{BC}$
In general $\vec{PQ} - \vec{PR} = \vec{RQ}$.

45

Mid Point

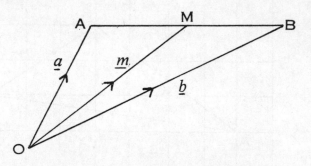

Let M be the mid point of $\overrightarrow{AB}$ then:—

$$\overrightarrow{AB} = \overrightarrow{OB} - \overrightarrow{OA}$$
$$= \underline{b} - \underline{a}$$

$$\overrightarrow{OM} = \overrightarrow{OA} + \tfrac{1}{2}\overrightarrow{AB}$$
$$= \underline{a} + \tfrac{1}{2}(\underline{b} - \underline{a})$$
$$= \underline{a} + \tfrac{1}{2}\underline{b} - \tfrac{1}{2}\underline{a}$$
$$= \tfrac{1}{2}\underline{a} + \tfrac{1}{2}\underline{b}$$
$$= \tfrac{1}{2}(\underline{a} + \underline{b})$$
$$= \tfrac{1}{2}(\overrightarrow{OA} + \overrightarrow{OB})$$

So far all the vectors have been 'free' vectors i.e. not 'tied' to any axes. Often it is more convenient to work with reference to fixed axes.

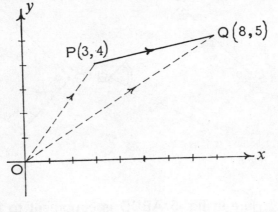

Fig 4

In fig. 4 given the point P(3, 4) we may express it as a directed line segment $\overrightarrow{OP}$ i.e. $\binom{3}{4}$ with reference to the origin. Again $\overrightarrow{OQ} = \binom{8}{5}$. Now $\overrightarrow{PQ}$ can be expressed in component form.

$$
\begin{aligned}
\overrightarrow{PQ} &= \overrightarrow{OQ} - \overrightarrow{OP} \\
&= \binom{8}{5} - \binom{3}{4} \\
&= \binom{8}{5} + \binom{-3}{-4} \\
&= \binom{5}{1}.
\end{aligned}
$$

47

Congruence

Two figures are said to be congruent if they are equal in all respects.

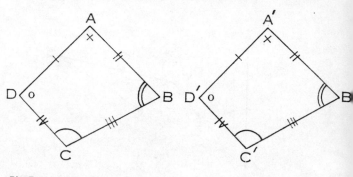

Fig 5

Here in fig. 5 ABCD is congruent to A'B'C'D' and vice versa.

∠A and ∠A' are called corresponding angles Similarly ∠C and ∠C' etc.

The sides AD and A'D' are called corresponding sides. Similarly AB and A'B' etc.

Similarity

Two figures are said to be similar if they have the same shape but not the same size i.e. each pair of corresponding angles are equal but not the corresponding sides.

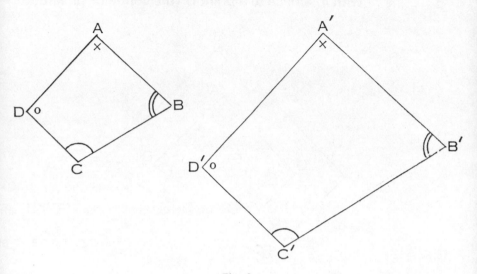

Fig. 6

In fig. 6 ABCD is similar to A'B'C'D'.
When two figures are similar their corresponding sides are in the same ratio, i.e.

$$\frac{AB}{A'B'} = \frac{BC}{B'C'} = \frac{CD}{C'D'} = \frac{DA}{D'A'} = k \text{ (in fig. 6 } k = \tfrac{1}{2})$$

From the ratios above we may write
AB = kA'B'; CD = kC'D' etc.
When similar figures have their corresponding sides parallel to each other we say the figures are similarly situated e.g. in fig. 6 ABCD and A'B'C'D' are similar and similarly situated.

Dilatations

A dilatation is a transformation of the plane, which will enlarge or reduce a figure to give an image figure, similar and similarly situated to the original figure. The magnitude of the enlargement or reduction is a factor k (the scale factor) as measured with reference to a point O (the centre of similitude).

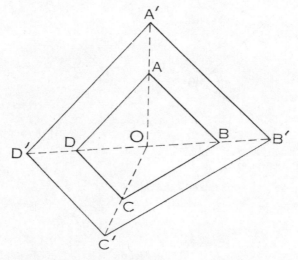

Fig. 7

In fig. 7 A'B'C'D' is a dilatation of ABCD i.e. ABCD has been mapped onto A'B'C'D' with O as the centre of similitude. In every dilatation the centre of similitude, a point and its image are all collinear, e.g. in fig. 7 O, A, A' are collinear, O, B, B' are collinear, O, C, C' etc. also under any dilatation:—

$$OA' = kOA \quad \text{and} \quad \overrightarrow{OA'} = k\overrightarrow{OA}$$

$$OB' = kOB \qquad \overrightarrow{OB'} = k\overrightarrow{OB} \quad \text{(in fig. 7 } k = 2\text{)}$$

$$OC' = kOC \qquad \overrightarrow{OC'} = k\overrightarrow{OC}$$

$$OD' = kOD \qquad \overrightarrow{OD'} = k\overrightarrow{OD}$$

Note that $\vec{OA'}$ and $\vec{OA}$ have the same sense and so k is +ve.

If k = 1 the figure and its image would be congruent.

If k > 1 the dilatation would be an enlargement.

If 0 < k < 1 the dilation would be a reduction.

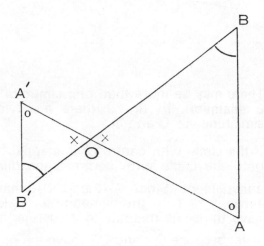

Fig. 8

In fig. 8 O, A, A' are collinear. Similarly O, B, B', so O is the centre of similitude and is invariant.

△A'OB' is a dilatation (here a reduction) of △AOB.

Here $\vec{OA}$ and $\vec{OA'}$ have opposite sense. Similarly $\vec{OB}$ and $\vec{OB'}$ have opposite sense, so the scale factor must be —ve i.e.

$\vec{OA'} = k\vec{OA}$, $\vec{OB'} = k\vec{OB}$ where k = —½.

We can indicate the dilatation by giving the centre of similitude and the scale factor thus [O, k]. In every dilatation a line and its image are parallel e.g. in fig. 8 $\vec{AB}$ is parallel to $\vec{A'B'}$; $\vec{AO}$ is parallel to $\vec{A'O}$ etc.

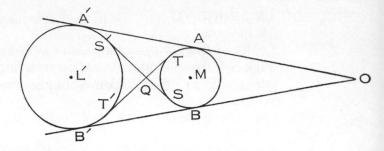

Fig. 9

There may be more than one centre of similitude for a dilatation. In fig. 9 there are two centres of similitude viz. O and Q.

If the circle with centre M is mapped under [O, k] onto the circle with centre L the dilatation is an enlargement. Since $\overrightarrow{OA}$ and $\overrightarrow{OA'}$ have the same sense, k > 1. If the dilatation is [Q, k] then k will have the same magnitude as before, but would be —ve, because $\overrightarrow{QT}$ and $\overrightarrow{QT'}$ have the opposite sense.

If the circle with centre L had been mapped onto the circle with centre M under the dilatations with centres O and Q, then we would have had a reduction, and the only difference from what has been said above is that the scale factor would be a +ve fraction (for centre O), or a —ve fraction (for centre Q). The centres of similitude of two circles can be found by drawing the exterior common tangents if k is to be +ve, or the interior common tangents if k is to be —ve. The point of intersection of the common tangents is the centre of similitude.

Reflection in a Line

Just as the reflection of a *man* in a mirror gives the image of a *man*.

The reflection of a *point* in a line gives the image of a *point*.

The reflection of a *line* in a line gives the image of a *line*.

The reflection of a *point* in a point gives the image of a *point*.

The reflection of a *line* in a point gives the image of a *line*.

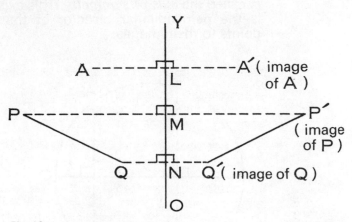

Fig. 10

Under a reflection in the line YO, A $<\!\!-\!\!-\!\!-\!\!>$ A'. (A goes to A', or A is mapped onto A' or vice versa).

Similarly P $<\!\!-\!\!-\!\!-\!\!>$ P', Q $<\!\!-\!\!-\!\!-\!\!>$ Q',

PQ $<\!\!-\!\!-\!\!-\!\!>$ P'Q'.

Note that the line YO is the perpendicular bisector of the lines joining points to their images.

Thus $\overrightarrow{AL} = \overrightarrow{LA'}$;

$\overrightarrow{PM} = \overrightarrow{MP'}$ etc.

If A is joined to P and Q, and A' is joined to P' and Q' then $\triangle$ APQ and $\triangle$ A'P'Q' are congruent.

53

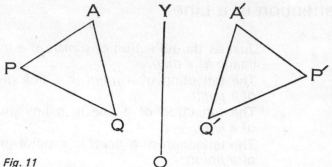

Fig. 11

When a figure and its image are congruent then the line (YO in fig. 11) in which the figure is reflected is called the *axis of symmetry*. This axis of symmetry is the perpendicular bisector of the lines joining points to their images.

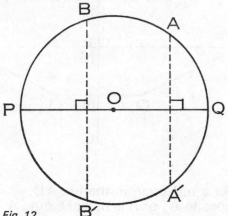

Fig. 12

The diameter PQ of the circle in fig. 12 is an axis of symmetry, and so bisects the chords AA' and BB' at right angles i.e. the chords are the lines joining the points on the circumference to their images under a reflection in the diameter. Also the semi-circle PBAQ is mapped onto PB'A'Q which is a congruent semi-circle. When a figure is mapped onto itself under a reflection in an axis of symmetry we say it has *bilateral symmetry*.

It should be noted that an axis of symmetry is mapped onto itself, and so any point on it will be mapped onto itself i.e. will be invariant. Any diameter of a circle is an axis of bilateral symmetry. Note that under a reflection in a line, lengths and sizes of angles are invariant.

Reflection in a Point

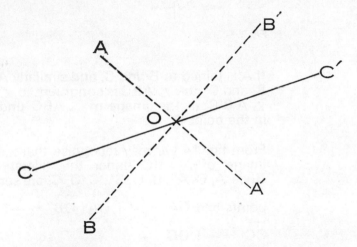

Fig. 13

In fig. 13 A, B, C are reflected in the point O. Note that O is the mid-point of the joins of the points and their images.

Thus $\overrightarrow{AO} = \overrightarrow{OA'}$,

$\overrightarrow{BO} = \overrightarrow{OB'}$,

$\overrightarrow{CO} = \overrightarrow{OC'}$

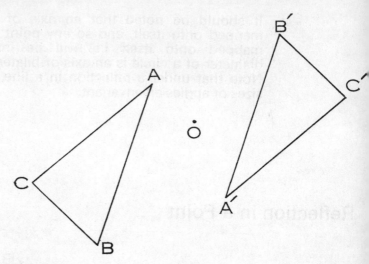

Fig. 14

If A is joined to B and C, and similarly A′ is joined to B′ and C′, the △ ABC is congruent to △ A′B′C′.
△ A′B′C′ is the image of △ ABC under reflection in the point O.

From fig. 14 we may recognise that △ A′B′C′ is the image of △ ABC under the dilatation [O, —1], since A, O, A′; B, O, B′; C, O, C′ are sets of collinear points and $\overrightarrow{OA'} = -1\,\overrightarrow{OA}$; $\overrightarrow{OB'} = -1\,\overrightarrow{OB}$; $\overrightarrow{OC'} = -1\,\overrightarrow{OC}$.

Rotation

A second look at reflection in a point reveals that the same end result is reached as when a point or a line is rotated about the point O through a half turn (180°).

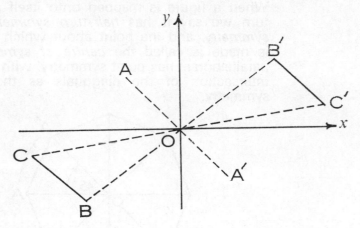

Fig. 15

Under a rotation of 180° about the origin

A<———>A';

B<———>B';

C<———>C';

BC<———>B'C'.

A half turn about O is then equivalent to a reflection in O, and both of these transformations are equivalent to the dilatation [O, —1].

Note that the △ COB is mapped onto △ C'OB' under a half-turn about O, and, as we have seen from the two equivalent transformations, the lengths of sides and angles are invariant by a rotation so △ COB is congruent to △ C'OB'.

Note that the point B is the intersection of the lines OB and CB, and that B' is the intersection of the lines OB' and C'B'. Thus the image of the intersection of two lines (i.e. the image of B) is the intersection of the images of the two lines (i.e. B').

57

In general the image of the point (a, b) under a rotation about the origin of 180° is the point (—a, —b).
Similarly the image of (a, b) under a reflection in O is (—a, —b).
Again the image of (a, b) under the dilatation [O, —1] is (—a, —b).

When a figure is mapped onto itself after a half-turn we say it has *half-turn symmetry* or *point symmetry,* and the point about which the rotation is made is called the *centre of symmetry* e.g. a parallelogram has point symmetry, with the point of intersection of the diagonals as the centre of symmetry.

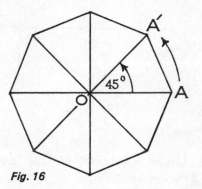

Fig. 16

This regular octagon in fig. 16 rotated anti-clockwise (taken to be the +ve direction of a rotation) about the point O through an angle of 45 ° will be mapped onto itself (A———>A'). When a figure is rotated about a point and thus is mapped onto itself we say the figure has *rotational symmetry*, and the point about which the figure is rotated is called the *centre of rotation.*
If figure 16 is rotated through 90 ° (i.e. 2 × 45 °) it is again mapped onto itself.
In fact the figure will be mapped onto itself after rotations of
1 × 45 °,
2 × 45 °,
3 × 45 °,
. . . . ,
n × 45 ° (n∈ + ve whole numbers).

The centre of rotation, O, is invariant. Since the octagon has 8 sides then A will be back where it started after $8 \times 45°$ rotations i.e. $\left(8 \times \dfrac{360}{8}\right)$ rotations.

In general a regular n-sided polygon is mapped onto itself by rotating about its centre through $\dfrac{360°}{n}$.

The number of rotations required to map any point A of the regular polygon onto itself again is called the order of the rotation.

The regular octagon has rotational symmetry of order 8.

A regular n-sided polygon has rotational symmetry of order n.

A regular polygon with an even number of sides has bilateral symmetry about any of its diagonals (i.e. the line joining opposite vertices).

A regular polygon with an odd number of sides will have no bilateral symmetry by joining vertices but it will have if a vertex is joined to the mid-point of the opposite side, as may be seen in the regular pentagon in fig. 17.

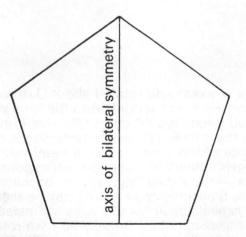

Fig. 17

Rotation through 90°

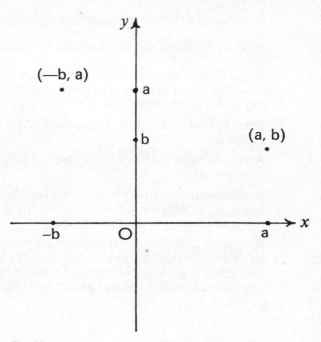

Fig. 18

If x, y-axes were rotated about O, through 90 ° then the +ve x-axis would go to the +ve y-axis, and the +ve y-axis would go to the —ve x-axis. Any value on the +ve x-axis would become the corresponding value of the +ve y-axis, and any value on the +ve y-axis would become the corresponding value on the —ve x-axis e.g. x = a becomes y = a and y = b becomes x = —b as may be seen from fig. 18. A similar result would arise if, instead of rotating the axes about O through 90°, we rotate a line or a point in the same manner, as is illustrated in fig. 19.

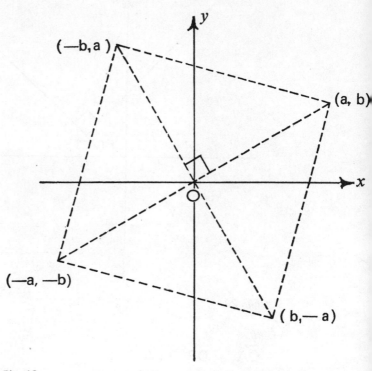

Fig. 19

By such a transformation the point (a, b) is mapped onto the point (—b, a) after the first ¼ turn about O. By a further ¼ turn about O the original point (a, b) would then be mapped onto (—a, —b) as we might expect from earlier considerations of $2 \times 90° = 180°$ or ½ turn transformation about 0. A further ¼ turn about O maps (—a, —b) onto (b, —a) as we might expect, since this is the result of a ¼ turn about O of the point (—b, a).

A rotation of 90° about a point is best dealt with by vector methods e.g. to find the image of A(8, 6) after ¼ turn about P(2, 1) as in Fig. 20.

61

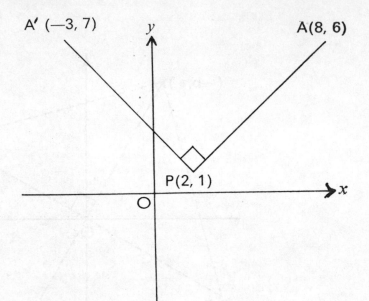

Fig. 20

$$\vec{OA} = \vec{OP} + \vec{PA}$$
$$= \binom{2}{1} + \binom{6}{5}$$

$$\vec{OA'} = \vec{OP} + \vec{PA'}$$
$$= \binom{2}{1} + \binom{-5}{6}$$
$$= \binom{-3}{7}$$

Thus the image of A is the point A′ (—3, 7).

Reflection in Axes

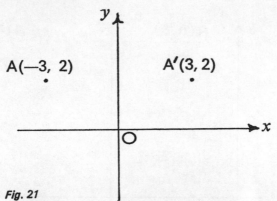

Fig. 21

A reflection in the y-axis of A(—3, 2) gives the image A'(3, 2) as in fig. 21. When a point is reflected in the y-axis the y co-ordinate is unaltered. The x co-ordinate is the same distance on one side of the y-axis as it is on the other, so that only the sign of the x co-ordinate changes. In general the image of (a, b) under a reflection in the y-axis is (—a, b).

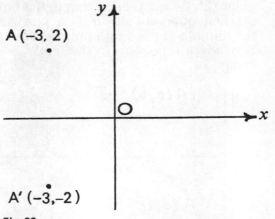

Fig. 22

In fig. 22 A' is the reflection of A in the x-axis so the x co-ordinate remains unaltered, but the sign of the y-co-ordinate is changed. In general the image of (a, b) under a reflection in the x-axis is (a, —b).

Reflection in a Line

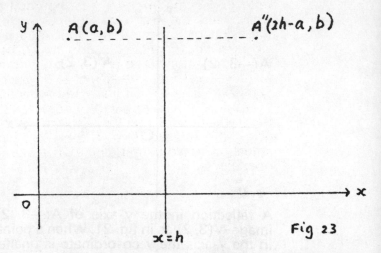

Fig 23

In fig. 23 A″ is the image of A under the reflection in the line x = h. Again the image is as far on one side of the line as A is on the other, and again the y-co-ordinate is unaltered. In general the image of (a, b) under a reflection in the line x = h is the point (2h — a, b). Note that in the figure (a, b) is in the first quadrant so 'a' is a positive number, but the formula applies to a point in any quadrant.

Similarly a reflection in the line y = k is as shown in fig. 24.

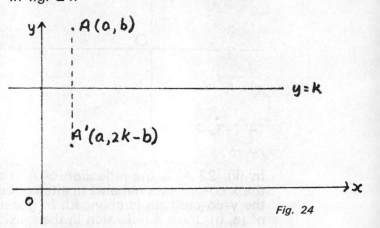

Fig. 24

In general the image of the point (a, b) under a reflection in the line y = k is the point (a, 2k — b). Combining the last two figures we can find the reflection of the point (a, b) in two perpendicular axes. In fig. 25 the reflection of (a, b) in the line x = h gives the image (2h —a, b), and the reflection of (2h — a, b) in the line y = k gives the image (2h — a, 2k — b). Similarly the reflection of (a, b) in the line y = k gives the image (a, 2k — b), and the reflection of (a, 2k — b) in the line x = h gives the image (2h — a, 2k — b). Note that the reflection in two perpendicular axes is a commutative operation.

Reflection in two Perpendicular Axes

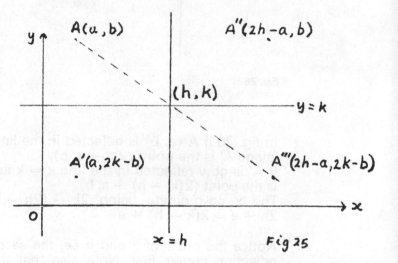

Fig 25

Note also that the reflection in the point (h, k) of the point (a, b) gives the image (2h — a, 2k — b) i.e. the reflection in x = h followed by a reflection in y = k is equivalent to a reflection in the point of intersection of the two perpendicular lines.

65

Reflection in Parallel Axes

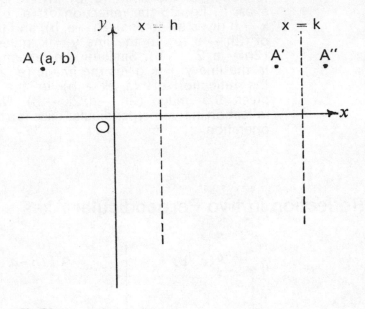

Fig. 26

In fig. 26 if A (a, b) is reflected in the line $x = h$, its image A′ is the point $(2h - a, b)$.
If A′ is now reflected in the line $x = k$ its image A″ is the point $(2(k - h) + a, b)$.
The x co-ordinate being $2k - (2h - a) = 2k - 2h + a = 2(k - h) + a$.

Notice the order of k and h *i.e.* the second axis of reflection comes first. Note also that the distance between A and A″ is twice the distance between the axes in which (a, b) is being reflected i.e. $|\overrightarrow{AA''}| = 2(k - h)$. Thus the result of a reflection in two parallel axes is equivalent to a simple translation of the point over a distance equal to twice the distance between the axes of reflection.

Alternatively by considering vectors as in fig. 27.

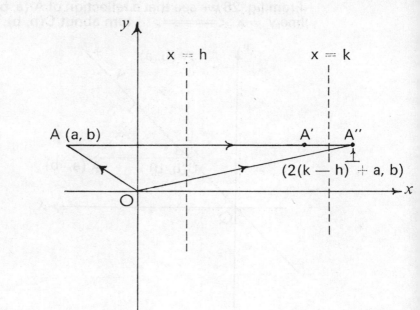

Fig. 27

$$\overrightarrow{OA} = \begin{pmatrix} a \\ b \end{pmatrix} \quad \overrightarrow{AA''} = \begin{pmatrix} 2(k-h) \\ 0 \end{pmatrix}$$

$$\begin{aligned}
\overrightarrow{OA''} &= \overrightarrow{OA} + \overrightarrow{AA''} \\
&= \begin{pmatrix} a \\ b \end{pmatrix} + \begin{pmatrix} 2(k-h) \\ 0 \end{pmatrix} \\
&= \begin{pmatrix} 2(k-h) + a \\ b \end{pmatrix}
\end{aligned}$$

i.e. A'' is the point $(2(k-h) + a, b)$.

REFLECTION IN THE LINE y = x

From fig. 28 we see that a reflection of A (a, b) in the line y = x <=====> ¼ turn about C(b, b).

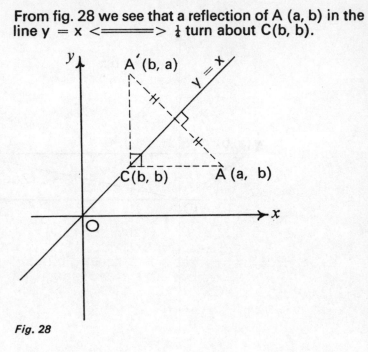

Fig. 28

$$\overrightarrow{OA} = \overrightarrow{OC} + \overrightarrow{CA}$$

$$= \binom{b}{b} + \binom{a-b}{b-b}$$

$$= \binom{b}{b} + \binom{a-b}{0}$$

$$\overrightarrow{OA'} = \overrightarrow{OC} + \overrightarrow{CA'}$$

$$= \binom{b}{b} + \binom{0}{a-b}$$

$$= \binom{b}{a}$$

Thus A' is the point (b, a).

Triangle Theorems

Medians

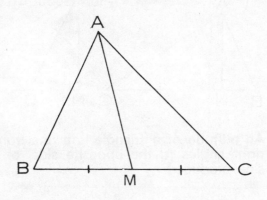

A line from a vertex of a triangle to the mid-point of the opposite side is called a median.
A median bisects the area of a triangle.

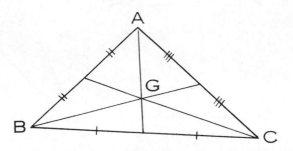

The medians of a triangle are concurrent. The point of concurrency, G, is called the centroid of the triangle and is the centre of gravity of the triangle.

Altitudes

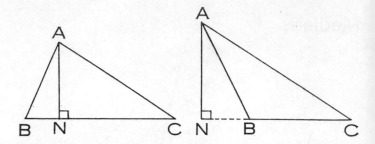

An altitude of a triangle is a line from a vertex at right angles to the opposite side, or the opposite side produced.

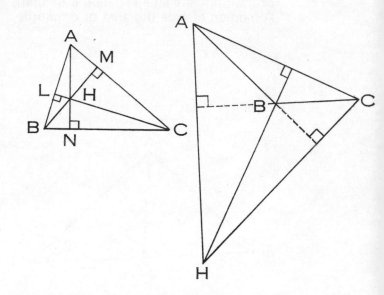

The altitudes of a triangle are concurrent. The point of concurrency, H, is called the orthocentre and may lie outside the triangle.

Angle Bisectors

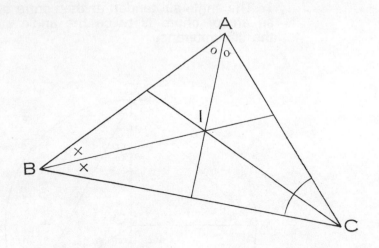

The internal bisectors of the angles of a triangle are concurrent. The point of concurrency, I, is called the incentre. With I as centre, a circle can be drawn inside the triangle such that the three sides are tangents to the circle.

Geometry Theorems

1. The angle subtended at the centre of a circle by an arc or chord is twice the angle subtended at the circumference.

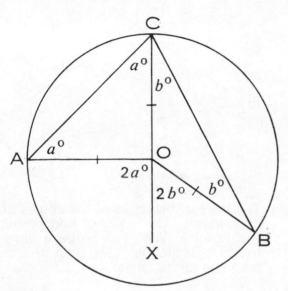

Fig. G. 1

In fig. G1 the minor (smaller) arc AB subtends ∠ AOB at the centre, and ∠ ACB at the circumference. CO is produced to X.

In △ AOC, ∠ ACO = ∠ CAO (opposite equal
                     = a°   sides; OC and OA
                                 are radii of the same
                                 circle)

∠ COA + a° + a° = 180° (The sum of the angles
                                  of a triangle)

But ∠ COA + ∠ AOX = 180° (COX is a straight line)
<=====> ∠ AOX = 2a°
Similarly ∠ BOX = 2b°
    Hence ∠ AOB = 2(a+b)°⎫ <=====>
    But ∠ ACB = (a + b)°⎭ *∠AOB = 2∠ACB.*

By using this property we can prove other theorems concerning the circle as follows.

2. An arc or chord of a circle subtends equal angles at the circumference.

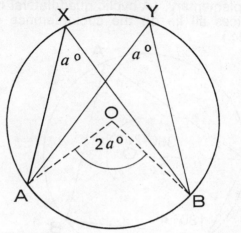

Fig. G. 2

From fig. G.2.

$$\angle AOB = 2 \angle AYB \brace \text{and } \angle AOB = 2 \angle AXB} \Longrightarrow \angle AYB = \angle AXB$$

3. The angle subtended at the circumference by a diameter is a right angle.

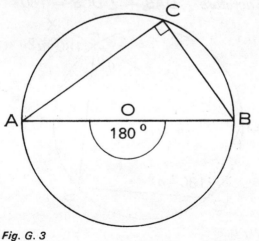

Fig. G. 3

From fig. G.3.
$$\left.\begin{array}{l}\angle \text{ AOB} = 180° \\ \text{and } \angle \text{ AOB} = 2 \angle \text{ ACB}\end{array}\right\} \Longrightarrow \angle \text{ ACB} = 90°$$

4. The opposite angles of a cyclic quadrilateral are supplementary. (A cyclic quadrilateral is one whose vertices all lie on the circumference of the same circle.)

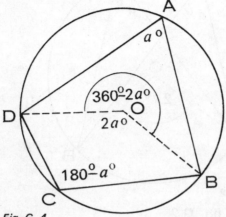

Fig. G. 4

In fig. G.4. the smaller angle DOB = 2 ∠ DAB
Also the greater angle DOB = 2 ∠ DCB
But the smaller ∠ DOB + the greater ∠ DOB = 360°
Therefore ∠ DAB + ∠ DCB = 180°

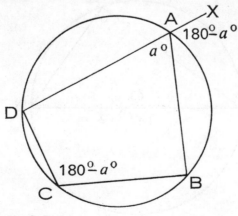

Fig. G. 5

74

5. In figure G.5. $\angle$ BAX is called the exterior angle the cyclic quadrilateral, ABCD. The exterior angle of of a cyclic quadrilateral is equal to the opposite interior angle. In fig. G.5. $\angle$ DCB is the opposite interior angle to which we refer.

From fig. G.5.

$$\left. \begin{array}{l} \angle \text{ DAB} + \angle \text{ DCB} = 180° \\ \angle \text{ DAB} + \angle \text{ BAX} = 180° \end{array} \right\} => \angle \text{ DCB} = \angle \text{ BAX}$$

The following deductions can be made from the preceding theorems.

1. If two angles subtended by a line are equal then the line is the chord of a circle and the two angles are subtended at the circumference of that circle i.e. the vertices of the two angles and the extremities of the line form the vertices of a cyclic quadrilateral.

2. If a line subtends a right angle then that line is the diameter of the circle which passes through the vertex of the right angle.

3. If the opposite angles of a quadrilateral are supplementary then the quadrilateral is cyclic.

4. If the exterior angle of a quadrilateral is equal to the opposite interior angle then that quadrilateral is cyclic.

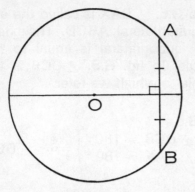

Theorem

A diameter at right angles to a chord bisects the chord.

Converse

A diameter which bisects a chord is at right angles to the chord.

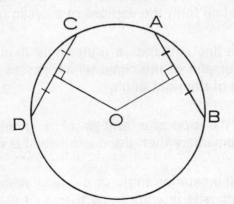

Theorem

If two chords are equidistant from the centre of a circle they are equal.

Converse

If two chords are equal they are equidistant from the centre.

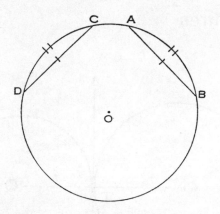

Theorem

If two chords are equal the arcs subtending them are equal.

Converse

If two arcs are equal the chords they subtend are equal.

Note

Problems involving distances from the centre, length of radius of the circle or lengths of chords are usually solved by the use of the theorem of Pythagoras since the line from the centre, the radius of the circle and half the chord form the sides of a right-angled triangle.

Sizes of angles involved in such problems may be found by direct application of trigonometric ratios.

Circles and Centres

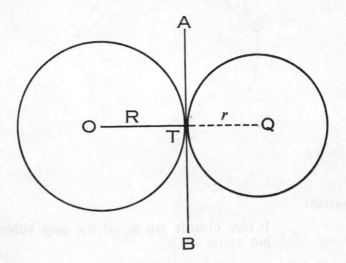

In the figure above, the line OQ is the line joining the centres of the two circles. Such a line is called the *line of centres*.

Notice that the line of centres goes through the point of contact, T, of the two circles.

Notice that the common tangent, AB, to the two circles cuts the line of centres at right angles.

Notice that the distance, d, between the centres of two circles touching externally is equal to the sum of the radii *i.e.* d = R + r.

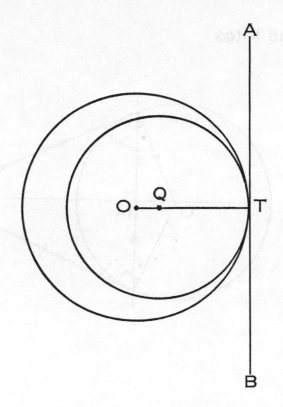

In this figure the circles touch internally but still the line of centres goes through the point of contact of the two circles, and again the common tangent meets this line at right angles. However this time the distance, d, between the two centres is the difference between the radii *i.e.* d = R — r.

Note that the angle between a radius and a tangent at the point of contact is a right angle.

The converse is also true i.e. if a radius of a circle meets a line at right angles at the circumference then that line is a tangent to the circle.

Circles and Kites

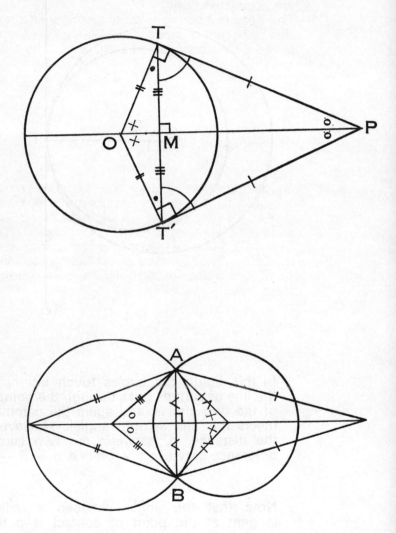

Above are two figures which turn up regularly in problems. Notice that all the quadrilaterals formed are kites.

The top figure shows the angle between a radius and tangent is a right angle. This will make the kite T PT'O a cyclic quadrilateral. (Opposite angles are supplementary.)
The bottom figure shows the normal kite shape and the 'delta-wing' kite shape. Notice that the common chord AB is bisected at right angles by the 'line of centres'.

Geometry Proofs and Deductions

Before attempting to prove or deduce what is required it is necessary to examine the basic properties of the figure given, e.g. point symmetry, rotational symmetry, line symmetry. Having done this we are in a better position to tackle the problem logically. The following types of arguments are common and you should familiarise yourself with them.

Symmetry in a Line

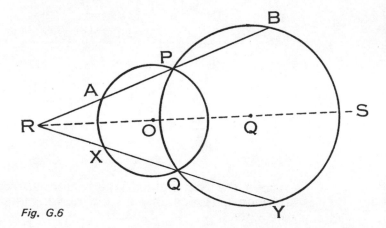

Fig. G.6

Given fig. G.6 prove i) PR = QR

ii) A is mapped onto X under a certain reflection.

iii) AB = XY

iv) AY = XB

Type 1) In fig. G.6 RS is an axis of symmetry for both circles.

Under reflection in RS $\quad$ P $<\!\!-\!\!-\!\!-\!\!>$ Q

$\qquad\qquad\qquad$ R $<\!\!-\!\!-\!\!-\!\!>$ R (R is invariant)

$\qquad\quad<\!\!=\!\!=\!\!=\!\!>$ PR $<\!\!-\!\!-\!\!-\!\!>$ QR

$\qquad\quad<\!\!=\!\!=\!\!=\!\!>$ PR = QR

Type 2) From (i) we may say that :—

the intersection of RP and the circle $<\!\!-\!\!-\!\!-\!\!-\!\!-\!\!>$ the intersection of RQ and the circle

$<\!\!=\!\!=\!\!=\!\!>$ A $<\!\!-\!\!-\!\!-\!\!>$ X.

Type 3) By the same argument as in type 2 :—

$\qquad$ B $<\!\!-\!\!-\!\!>$ Y $\Big\}$ $\quad<\!\!=\!\!=\!\!>$ $\quad$ AB $<\!\!-\!\!-\!\!>$ YX

but $\qquad$ A $<\!\!-\!\!-\!\!>$ X $\qquad\qquad\qquad$ AB $\quad=\quad$ XY

Also $\quad$ AY $<\!\!-\!\!-\!\!>$ XB

$<\!\!=\!\!>$ AY $\quad=\quad$ XB

Rotational Symmetry

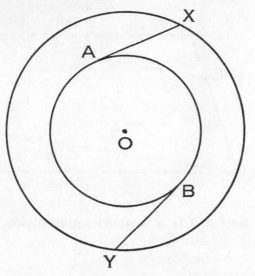

Given two concentric circles with centre O with AX and BY tangents to the inner circle at A and B respectively, and meeting the outer circle at X and Y respectively, show that AX = BY.

Type 4) Under a rotation about O.

OA——————> OB <═════> A————————>B.

The line BY will lie on the line AX (because $\angle OBY = \angle OAX = 90°$)

<═════> the intersection of BY and the outer circle————> the intersection of AX and the outer circle

<═════> X————> Y

Now A————>B⎫
and X————>Y⎭ <═> AX————>BY <═> AX = BY

83

Point Symmetry

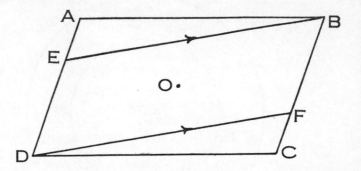

Given ABCD is a parallelogram with BE parallel to FD.

By rotation prove $\triangle$ ABE is congruent to $\triangle$ CDF.

Under $\frac{1}{2}$ turn (rotation of 180°) about O (the point of intersection of the diagonals of ABCD).

$$\left.\begin{array}{l} A <\longrightarrow> C \\ B <\longrightarrow> D \end{array}\right\} <===> \left\{\begin{array}{l} AB <\longrightarrow> CD \\ AD <\longrightarrow> CB \end{array}\right.$$

The line BE will lie on the line DF
the intersection of BE and the line AD $<\longrightarrow>$
the intersection of DF and the line CB
$<==>$ E $<\longrightarrow>$ F

Now $\left.\begin{array}{l} A <\longrightarrow> C \\ E <\longrightarrow> F \end{array}\right\} <=> AE <\longrightarrow> CF <=> AE = CF$

$\left.\begin{array}{l} B <\longrightarrow> D \\ E <\longrightarrow> F \end{array}\right\} <=> BE <\longrightarrow> DF <=> BE = DF$

AB = DC (opposite sides of parallelogram ABCD)
$\therefore$ $\triangle$ ABE is congruent to $\triangle$ CDF (3 sides equal).

Points to Note

a) The shape of the argument.
e.g. if under a reflection (in a point or a line)

A <————> B
C <————> D   (Notice double arrow)

then line segment AC <————> line segment BD
and AC = BD.

b) Using the facts above again.

A <————> B
C <————> D

then line segment AD <————> line segment BC
and AD = BC.

c) In contrast with the two equal pairs of line segments found in (a) and (b) only one pair of equal line segments can be found when the image point is from a rotation other than through 180°.
e.g. If under a rotation about a point

A ————> B   (Notice single arrow)
C ————> D

then line segment AC ————> line segment BD
and AC = BD.

d) Notice that the image of an intersection is the intersection of the image. This fact arises regularly in proofs and deductions, and often is the key point required to complete the argument. Look over the notes on this point again and revise the examples which use this fact.

e) Any point lying on the axis of symmetry is invariant under reflection in that axis. Similarly in reflection in a point, that point is invariant.

f) Under rotation about a point, that point is invariant.

Trigonometry

The Cartesian diagram is divided into four **quadrants** as in fig. 1.

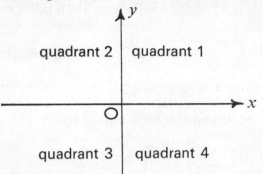

Fig. 1

Any point in quadrant 1 will have (+ve, +ve) co-ordinates.
Any point in quadrant 2 will have (—ve, +ve) co-ordinates.
Any point in quadrant 3 will have (—ve, —ve) co-ordinates.
Any point in quadrant 4 will have (+ve, —ve) co-ordinates.
When a point P lies in quadrant 1 then the line segment $\overrightarrow{OP}$ makes an acute angle with the +ve direction of the x-axis, e.g. in fig. 2. $\overrightarrow{OP}$ makes an angle a ° with the +ve direction of the x-axis.

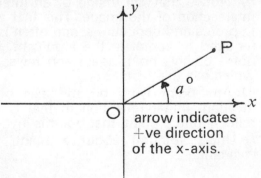

Fig. 2

$\angle$ XOP is formed by a rotation from $\overrightarrow{OX}$ to $\overrightarrow{OP}$

When this rotation is counter-clockwise the angle is considered +ve e.g. in fig. 2. $\angle$ XOP = +a°.

When the rotation is clockwise the angle thus formed is considered —ve e.g. in fig. 3 $\overrightarrow{OP}$ has rotated about O through —a°, again from the +ve direction of the x-axis. Note that the +ve angle $\angle$ XOP in fig. 3 is (360 — a)°.

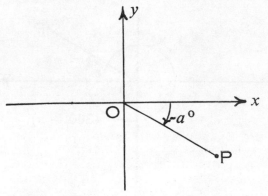

Fig. 3

Fig. 4. shows that when P is in quadrant 2 then $\overrightarrow{OP}$ makes an obtuse angle with the +ve direction of the x-axis.

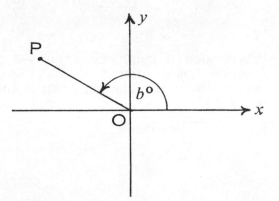

Fig. 4

87

As the line segment $\overrightarrow{OP}$ sweeps about the origin it eventually covers the same ground again. In fig. 5 $\overrightarrow{OP}$ has swept through $(360 + a)°$ and so is back in quadrant 1 and in the same position as when it had swept out only $+a°$.

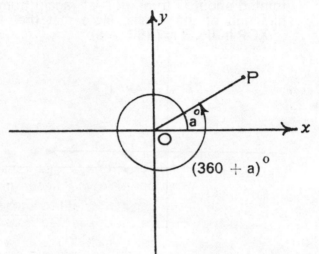

Fig. 5

Similarly if $\overrightarrow{OP}$ had swept through $(2 \times 360 + a)°$
       or $(3 \times 360 + a)°$
       or $(7 \times 360 + a)°$
       or $(n \times 360 + a)°$

When such a sequence arises we say that the function has a *period* of $360°$ i.e. the relative position from the starting point is the same after every $360°$.

Sine, Cosine, Tangent

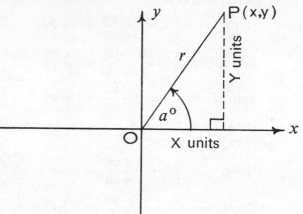

Fig. 6

From fig. 6.

$$\text{Sin } a° = \frac{y}{r} = \frac{\text{number of units P lies from the x-axis}}{\text{number of units P lies from the origin}}$$

$$\text{Cos } a° = \frac{x}{r} = \frac{\text{number of units P lies from the y-axis}}{\text{number of units P lies from the origin}}$$

$$\text{Tan } a° = \frac{y}{x} = \frac{\text{number of units P lies from the x-axis}}{\text{number of units P lies from the y-axis}}$$

Note:

a) $\dfrac{\text{Sin } a°}{\text{Cos } a°} = \dfrac{y/r}{x/r} = \dfrac{y}{x} = \text{Tan } a°$

b) $\text{Sin}^2 a° + \text{Cos}^2 a° = \left(\dfrac{y}{r}\right)^2 + \left(\dfrac{x}{r}\right)^2$

$$= \frac{y^2}{r^2} + \frac{x^2}{r^2} = \frac{x^2 + y^2}{r^2}$$

but, by the theorem of Pythagoras, $x^2 + y^2 = r^2$,

so $\text{Sin}^2 a° + \text{Cos}^2 a° = \dfrac{r^2}{r^2} = 1$.

Notice that Sin a° or Cos a° must always be less than or equal numerically to unity i.e. $-1 \leqslant \text{Sin } a°$, Cos a° $\leqslant 1$, because r is the hypoteneuse of the right angled triangle from which we read the sine and cosine, and so r is always greater or equal to x or y numerically i.e. $|x| \leqslant r$ and $|y| \leqslant r$.

When P is in quadrant 1 its co-ordinates are (+ve, +ve) and $0 < a < 90$.

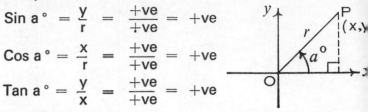

$$\text{Sin } a° = \frac{y}{r} = \frac{+ve}{+ve} = +ve$$

$$\text{Cos } a° = \frac{x}{r} = \frac{+ve}{+ve} = +ve$$

$$\text{Tan } a° = \frac{y}{x} = \frac{+ve}{+ve} = +ve$$

When P is in quadrant 2 its co-ordinates are (—ve, +ve), $90 < b < 180$,

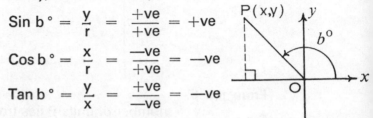

$$\text{Sin } b° = \frac{y}{r} = \frac{+ve}{+ve} = +ve$$

$$\text{Cos } b° = \frac{x}{r} = \frac{-ve}{+ve} = -ve$$

$$\text{Tan } b° = \frac{y}{x} = \frac{+ve}{-ve} = -ve$$

When P is in quadrant 3 its co-ordinates are (—ve, —ve), $180 < c < 270$.

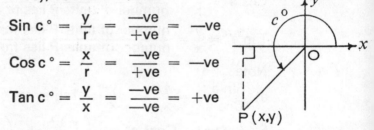

$$\text{Sin } c° = \frac{y}{r} = \frac{-ve}{+ve} = -ve$$

$$\text{Cos } c° = \frac{x}{r} = \frac{-ve}{+ve} = -ve$$

$$\text{Tan } c° = \frac{y}{x} = \frac{-ve}{-ve} = +ve$$

When P is in quadrant 4 its co-ordinates are (+ve, —ve), $270 < d < 360$.

$$\text{Sin } d° = \frac{y}{r} = \frac{-ve}{+ve} = -ve$$

$$\text{Cos } d° = \frac{x}{r} = \frac{+ve}{+ve} = +ve$$

$$\text{Tan } d° = \frac{y}{x} = \frac{-ve}{+ve} = -ve$$

All the above results are easily recalled by the mnemonic CAST as in fig 7.

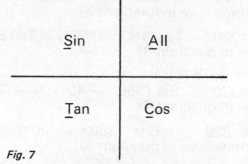

Fig. 7

The letters in the quadrants are the initials of the ratios which are +ve when the point P lies in that quadrant. Notice that the word CAST is read in a counter clockwise direction as we would read a +ve angle, and that the letter 'A', the first letter in the alphabet, goes into the first quadrant. From CAST we can read the Sines, Cosines and Tangents of all angles from the tables.

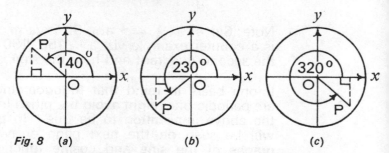

Fig. 8 (a) (b) (c)

From fig. 8 (a)
Sin 140° = Sin (180° − 40°) = Sin 40° (Sine +ve in quadrant 2)

Cos 140° = Cos (180° − 40°) = −Cos 40° (Cosine −ve in quadrant 2)

Tan 140° = Tan (180° − 40°) = −Tan 40° (Tangent −ve in quadrant 2)

From fig. 8 (b).
Sine 230° = Sin (180° + 50°) = −Sin 50° (Sine −ve in quadrant 3)

Cos 230° = Cos (180° + 50°) = —Cos 50°
(Cosine —ve in quadrant 3)

Tan 230° = Tan (180° + 50°) = Tan 50° (Tangent +ve in quadrant 3)

From fig. 8 (c).
Sin 320° = Sin (360° — 40°) = —Sin 40° (Sine —ve in quadrant 4)

Cos 320° = Cos (360° — 40°) = Cos 40°
(Cosine +ve in quadrant 4)

Tan 320° = Tan (360° — 40°) = —Tan 40°
(Tangent —ve in quadrant 4)

Notice that all angles XOP are formed by dropping a perpendicular from P to the x-axis thus forming right-angled triangles from which we read off the sines, cosines and tangents according to our definitions $\frac{y}{r}, \frac{x}{r}, \frac{y}{x}$ respectively.

Note: Sin a° = $\frac{1}{2}$ => a = 30 is *false*, since there is a counter example viz. a = 150. 150° puts P in the second quadrant and here the sine is also +ve.

If one bears in mind that trigonometric functions are periodic one might avoid the pitfall in supposing the above implication to be true. In particular, it will be seen on the next page or two that the graphs of the sine and cosine functions display symmetry, which is again an aid in dealing with such an example as this.

Polar Co-ordinates

To fix a point on a Cartesian diagram we require two co-ordinates e.g. P (x, y). By this method we go along the x-axis and y-axis the stated number of units. Alternatively we can fix the point P by stating its distance from the origin, r, and the angle, θ°, which $\overrightarrow{OP}$ makes with the +ve direction of the x-axis. This would give the co-ordinates of the point P (r, θ°), which are known as the polar co-ordinates.

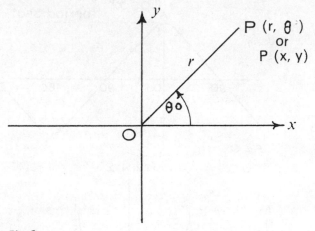

Fig. 9

It is possible to express one form in terms of the other e.g.

$$\text{Sin } \theta° = \frac{y}{r} \quad <\!\!=\!\!=\!\!=\!\!> \quad y = r \text{ Sin } \theta°$$

$$\text{Cos } \theta° = \frac{x}{r} \quad <\!\!=\!\!=\!\!=\!\!> \quad x = r \text{ Cos } \theta°$$

Thus the point P in fig. 9 may be written
P (r Cos θ°, r Sin θ°).

Note again $\text{Tan } \theta° = \dfrac{y}{x} = \dfrac{r \text{ Sin } \theta°}{r \text{ Cos } \theta°} = \dfrac{\text{Sin } \theta°}{\text{Cos } \theta°}$

Graphs

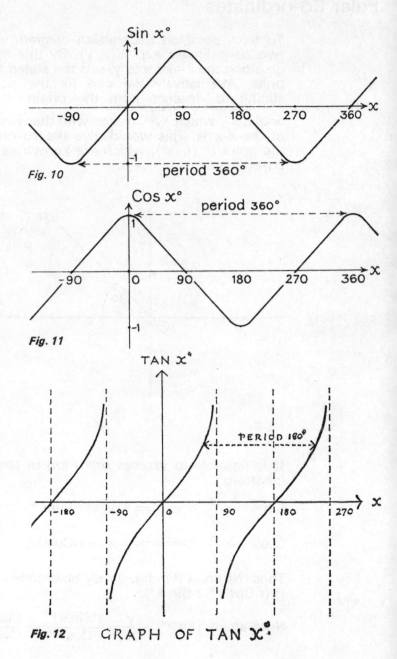

Fig. 10

period 360°

Fig. 11

Cos x°   period 360°

Fig. 12   GRAPH OF TAN x°

Fig. 10 shows that Sin x ° has a maximum value of 1 and a minimum value of —1. It also shows that Sin x° is a periodic function, since the pattern is repeated every 360 °.

Fig. 11 shows that Cos x ° has a maximum value of 1 and a minimum value of —1. It also shows that Cos x ° is a periodic function since the pattern is repeated every 360 °.

For both Sin x° and Cos x° the domain is the set of angles, and the range —1 $\leqslant$ f(x) $\leqslant$ 1, f(x) $\in$ R.

From fig. 10 and fig. 11 we see that the graph of Sin x ° is similar to the graph of Cos x ° but translated 90° along the axis of the domain, so Cos x° = Sin (x ° + 90 °).

Fig. 12 shows the graph of Tan x °. Notice that Tan x ° has no maximum and no minimum value. There is no point in the range on which to map the points in the domain where the values are the odd multiples of 90 ° e.g. 1 $\times$ 90 °, 3 $\times$ 90 °, (2n — 1) $\times$ 90 °.

The period of Tan x ° is half the period of sin x ° and cos x ° i.e. the period of Tan x ° is 180 °, since the pattern is repeated every 180 °.

Evaluating Ratios of certain Angles without Tables

Some ratios can be evaluated for certain angles without tables by the use of certain triangles.

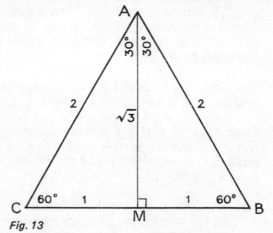

Fig. 13

To evaluate Sine, Cosine, Tangent of 30° or 60° we use an equilateral triangle with sides 2 units, as in fig 13. By drawing an altitude from A to BC we bisect BC. By the theorem of Pythagoras AM = $\sqrt{3}$. Now following the definitions we find Sin 60° = $\sqrt{\frac{3}{2}}$, Cos 60° = $\frac{1}{2}$, Tan 60° = $\sqrt{3}$. Sin 30° = $\frac{1}{2}$, Cos 30° = $\sqrt{\frac{3}{2}}$, Tan 30° = $\frac{1}{\sqrt{3}}$

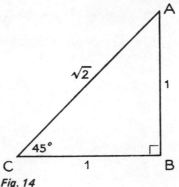

Fig. 14

To calculate Sine, Cosine and Tangent of 45° we use an isosceles, right-angled triangle, whose equal sides are 1 unit in length. Again by the theorem of Pythagoras the hypoteneuse is $\sqrt{2}$ units in length as in fig 14. Now Sin 45° = $\frac{1}{\sqrt{2}}$, Cos 45° = $\frac{1}{\sqrt{2}}$, Tan 45° = 1.

Solving Triangles

When we find the sizes of the sides and angles of a triangle we say we solve the triangle. Angles are named by the use of capital letters, and sides are named by the use of lower case letters to correspond with the angles opposite them as in fig. 15.

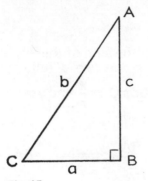

Fig. 15

Remember: The greatest side is opposite the greatest angle.
The smallest side is opposite the smallest angle.
The middle-sized side is opposite the middle sized angle.
The sum of any 2 sides of a triangle must always be greater than the 3rd side (triangular inequality).

When the size of any side or angle has been found you should check that the result is reasonable by considering its magnitude relative to the other given magnitudes.

Method

Consider the sides about the right angle as x, y, and the hypoteneuse as r. Place one of the vertices of the acute angles on the origin of a Cartesian diagram and a side along the x-axis e.g. solve $\triangle$ ABC when $\angle$ B = 90°, $\angle$ C = 30°, b = 2.

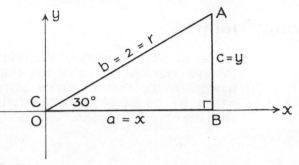

Fig. 16

$\angle$A = 60° (from the sum of the angles of a triangle.) Since r is known we use a ratio which includes

e.g. Sin C = $\dfrac{y}{r}$, Cos C = $\dfrac{x}{r}$ and so obtain the magnitudes of BA and CB.

$$\text{Sin } 30° = \frac{y}{r} = \frac{BA}{2} \Longleftrightarrow BA = 2 \text{ Sin } 30° = 1$$
$$\text{Cos } 30° = \frac{x}{r} = \frac{CB}{2} \Longleftrightarrow CB = 2 \text{ Cos } 30° = \sqrt{3}$$

Similarly, if we are required to find $\angle$C in fig. 16 given b = 2, c = 1, $\angle$B = 90°

$$\text{Sin C}° = \frac{y}{r} = \frac{BA}{AC} = \frac{1}{2} \Longrightarrow C = 30°$$

Or find $\angle$C given c =1, a = $\sqrt{3}$. Here we are given values for x, y so we use ratios that do not include r *i.e.*

$$\text{Tan C}° = \frac{y}{x} = \frac{BA}{BC} = \frac{1}{\sqrt{3}} \Longrightarrow \angle C = 30°$$

When the triangle which we are attempting to solve is not right-angled we may be able to use the Sine Rule i.e. $\dfrac{a}{Sin\ A} = \dfrac{b}{Sin\ B} = \dfrac{c}{Sin\ C}$

As before, we consider one vertex to lie on the origin and one side to lie along the x-axis.

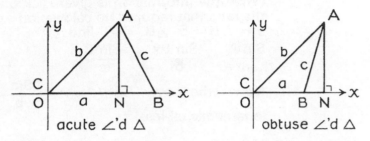

Fig. 17

Fig. 17 shows the two possibilities which could arise. In both cases we can draw a perpendicular from A to the x-axis. Thus :—

AN = AC Sin C = b Sin C

AN = AB Sin (180 ° — B) = c Sin B.

It follows then that b Sin C = c Sin B $<\!\!=\!\!=\!\!=\!\!>$

$\dfrac{b}{Sin\ B} = \dfrac{c}{Sin\ C}$

Thus we can dispense with AN and the Cartesian diagram, since $\dfrac{b}{Sin\ B} = \dfrac{c}{Sin\ C}$ makes use of only the sides and angles of the original triangle ABC.

Similarly by placing another vertex (A, say) on the origin we could obtain the result $\dfrac{a}{Sin\ A} = \dfrac{b}{Sin\ B}$

$= \dfrac{c}{Sin\ C}$ and with these results we have the Sine Rule.

If the given triangle is right-angled then we shoul
not use the Sine Rule but employ the method show
in the earlier section.

If a slide rule is not being used for calculations the
the following form of the Sine Rule is more con
venient when an angle is to be found:

$$\frac{Sin\ A}{a} = \frac{Sin\ B}{b} = \frac{Sin\ C}{c}$$

When the information is given, tick off each part c
the ratio that requires no calculation e.g.
$a = 7, b = 5, \angle B = 48°$ find $\angle A$.

$$\frac{Sin\ A}{a^v} = \frac{Sin\ B^v}{b^v} = \frac{Sin\ C}{c}$$

Now write down Sin A = $\frac{a\ Sin\ B,}{b}$ if usin

logarithmic tables.

Unless we are given an angle and the side opposit
this angle we can not use the Sine Rule. If this is th
case then we may use the Cosine Rule.

Cosine Rule

If we wish to calculate the magnitude of a side we use the form $a^2 = b^2 + c^2 - 2bc \cos A$. If we wish to calculate an angle, then, by changing the subject of this formula, we have $\cos A = \dfrac{b^2 + c^2 - a^2}{2bc}$.

As before we let a vertex lie on the origin, and a side lie along the x-axis.

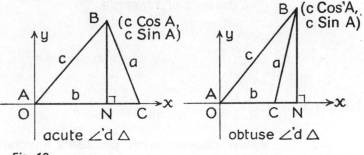

acute $\angle$'d $\triangle$ obtuse $\angle$'d $\triangle$

Fig. 18

Fig. 18 shows the two possibilities that could arise. In both cases we can draw a perpendicular from B to the x-axis.

Thus $AN = c \cos A \Longleftrightarrow x_B = c \cos A$
$BN = c \sin A \Longleftrightarrow y_B = c \sin A$

Now by the distance formula
$$BC^2 = (x_C - x_B)^2 + (y_C - y_B)^2$$
$\Longleftrightarrow a^2 = (b - c \cos A)^2 + (0 - c \sin A)^2$
$\Longleftrightarrow a^2 = b^2 - 2bc \cos A + c^2 \cos^2 A + c^2 \sin^2 A$
$\Longleftrightarrow a^2 = b^2 - 2bc \cos A + c^2 (\cos^2 A + \sin^2 A)$
$\Longleftrightarrow a^2 = b^2 + c^2 - 2bc \cos A.$

If the angle we are using in the formula is obtuse then we proceed as follows given $\angle B = 130°$.
$$b^2 = a^2 + c^2 - 2ac \cos 130°$$
$\Longleftrightarrow b^2 = a^2 + c^2 - 2ac (-\cos 50°)$
$\Longleftrightarrow b^2 = a^2 + c^2 + 2ac \cos 50°.$

101

When solving triangles start by finding the smalle[r]
angles. The third angle may be found from the sum
of the angles of a triangle, thus obviating the
necessity to remember that Cos (obtuse angle) is
—ve, which might be overlooked in the midst o[f]
other working.

From fig. 18 the area of $\triangle ABC = \frac{1}{2}AC. BN =$
$\frac{1}{2}bc$ Sin A.

Any two sides and the Sine of the included angle wi[ll]
give the area of a triangle.

Bearings

When dealing with problems involving compass
bearings, first draw a NS line and choose a poin[t]
from which to start as in fig. 19.

Fig. 19

There could be information giving a point B due east of A, or C due north of A. These are dealt with as in fig. 20.

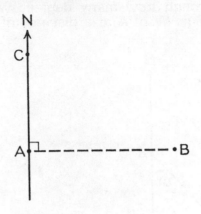

Fig. 20

Other bearings may be given on the form a ° where $0 < a < 360$. In this case we draw fig. 19 and then face north and turn a ° clockwise e.g. B is on a bearing a ° from A at a distance of c miles as in fig. 21.

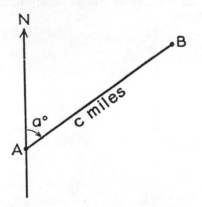

Fig. 21

Bearings could be given in the form N a° E or N a° W or S a° E or S a° W. In this case the first direction tells us which way to face, and the second direction tells us which way to turn, leaving a° to tell us through how many degrees we must turn e.g. B is S a° W of A at a distance of c miles (see fig. 22).

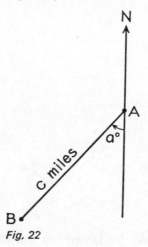

Fig, 22

From such bearings one could construct a right angled triangle by drawing a perpendicular from B to the NS line, and so utilise the properties of such a triangle to calculate other distances and angles in the figure.

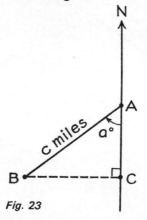

Fig. 23

In fig. 23 we may calculate how far B lies west of A by finding the length of BC i.e. BC = c Sin a °.
Again we may calculate how far B lies south of A by finding the length AC i.e. AC = c Cos a°.
Sometimes the problem is more involved, e.g. B is a bearing N a° W of A at a distance of c miles, and C is on a bearing S b ° E of B. First draw a NS line and choose a starting point A, then plot B as in fig. 24.

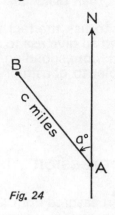

Fig. 24

Now treat the second part as the first, taking B as the starting point, and drawing a new NS line through B as in fig. 25.

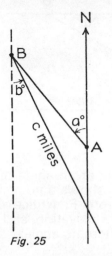

Fig. 25

Notice that we do not know exactly where C is, since we have been given no distance from any other point. Exactly where it lies will be indicated by the example. If we are required to find how far C will be from A when it is due S of A, then C will be the point where the NS line intersects BC.

When a figure like fig. 25 has been constructed then one may employ the Sine Rule or Cosine Rule in order to calculate other distances and angles.

One should remember to use the fact that the two NS lines are parallel, and so give rise to equal angles such as alternate and corresponding angles etc. Again it might be simpler to construct right-angled triangles as before.

Angles of Elevation and Depression

Angles of *elevation* are measured from the *horizontal* line *up*. Angles of *depression* are measured from a line parallel to the *horizontal* line *down*.

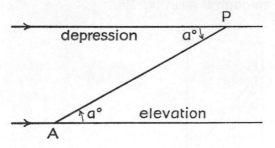

Fig. 26

Fig. 26 shows a point P at an angle of elevation of a° from A. Fig. 26 also shows a point A at an angle of depression of a° from P. Notice that between two points, the angle of elevation, and the angle of depression are equal alternate angles.

When dealing with a circle on a horizontal plane and a point not in that plane, the circle is best represented by an ellipse as in fig. 27.

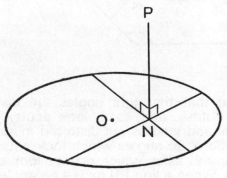

Fig. 27

The distance between a point and a plane is taken to be the shortest distance i.e. PN is the distance between the point P and the circle with centre O, and N is a point on the plane.

If a line, e.g. PN in fig. 27, is perpendicular to two lines in the plane going through N, then PN is perpendicular to all the lines in the plane going through N.

When dealing with a rectangle as the horizontal plane it is conveniently represented by a parallelogram as in fig. 28, and when dealing with a square in the horizontal plane one should draw a rhombus.

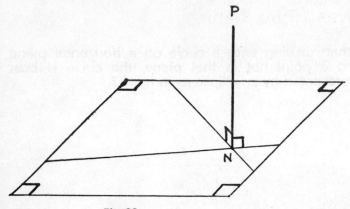

Fig. 28

Notice that the right angles are distorted. Some look obtuse, and some look acute. Similarly all angles and lengths are distorted in such 3D representations, so angles which look equal may not be equal, and sides which do not look equal may be equal. When a line PN makes an angle with a plane we take the angle to be the smallest one. As shown in fig. 29 ∠ PNX is the angle between PN and the plane ABCD i.e. ∠ PNX is the angle between PN and the projection of PN (i.e. NX) on the plane.

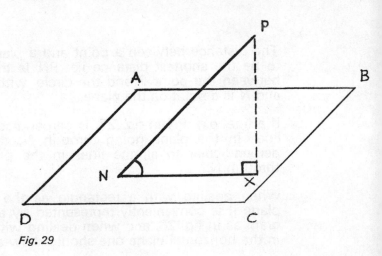

Fig. 29

Angle Between Two Planes

When two planes intersect, as in fig. 30, they do so in a line i.e. PQRS and ABCD intersect in the line XY.

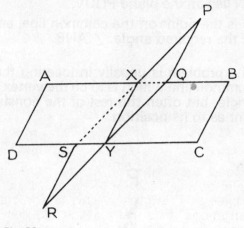

Fig. 30

The angle between two planes is taken to be the angle between two lines, one on each plane, and at right angles to the common line between the planes e.g. in fig. 31 ∠ ANB is the angle between the planes PQRS and PQUV.

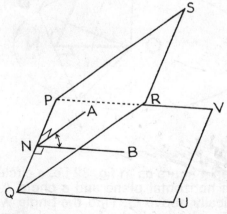

Fig. 31

Notice AN is drawn perpendicular to the common line PQ.

BN is drawn perpendicular to the common line PQ.

AN lies on the plane PQRS.

BN lies on the plane PQUV.

N is the point on the common line, and is the vertex of the required angle, $\angle$ ANB.

The problem is usually in locating the point on the common line which is to be the vertex of the required angle, but often the rest of the construction gives a hint as to its position.

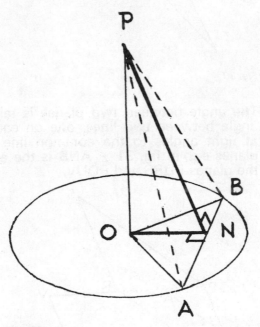

Fig. 32

Given a figure as in fig. 32 i.e. a circle with centre O on a horizontal plane and a chord AB. P is a point vertically above O. Find the angle which the plane PAB makes with the horizontal. The best approach is first to locate the common line between the two planes, then, using this line as a base, look for a possible isosceles or equilateral triangle with the common line as its base. When dealing with a figure such as fig. 32, we may use the fact that a line at right angles to a chord from the centre of the circle bisects the chord. The radii can give us the equal sides of an isosceles triangle. Remember too, that the altitude of an isosceles triangle or equilateral triangle will bisect the base at right angles. In fig. 32 $\angle$PNO is the angle between the circle and the plane PAB.

Space Diagonals

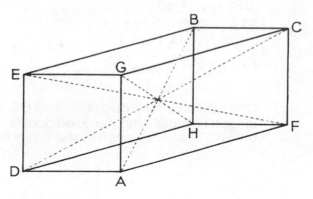

Fig. 33

Fig. 33 shows a cuboid with its 4 space diagonals. AB = CD = EF = GH. The lengths of these space diagonals can be calculated by using the theorem of Pythagoras twice over.

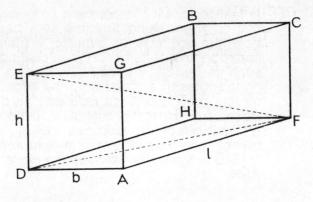

Fig. 34

In fig. 34 triangles EDF and ADF are right-angled.

$$DF^2 = l^2 + b^2$$
$$EF^2 = DF^2 + h^2$$
$$= l^2 + b^2 + h^2$$
$$EF = \sqrt{l^2 + b^2 + h^2}$$

DF is called a face diagonal, and DG is another face diagonal. Note that face diagonals are not equal unless $l = b = h$ *i.e.* when the figure is a cube.

3D Co-ordinates

To locate a point in two dimensions, e.g. on a Cartesian diagram, we require two co-ordinates i.e. x, y or r, $\theta°$. To locate a point in three dimensions we require three co-ordinates x, y, z. Figure 35 shows the point P(x, y, z) i.e. (2, 1, 4). P is 2 units along the x-axis, 1 unit along the y-axis, and 4 units along the z-axis.

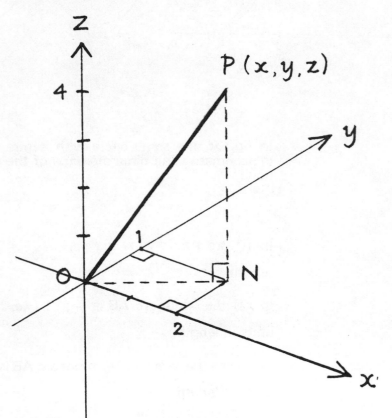

Fig. 35

Note that the distance OP is found by the same method as was used to find the length of space diagonals i.e. $OP = \sqrt{x^2 + y^2 + z^2} = \sqrt{2^2 + 1^2 + 4^2} = \sqrt{21}$.

Length of an Arc

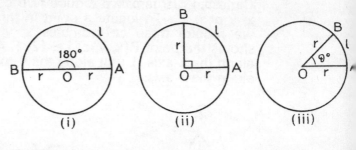

Fig. 36

In fig. 36 the same circle with centre O is drawn three times. The circumference of the circle is 2π units.

In (i) arc AB is $\frac{1}{2} \times 2\pi r$ *i.e.* $\frac{180}{360} \times 2\pi r$ units in length.

In (ii) the minor arc AB is $\frac{1}{4} \times 2\pi r$ *i.e.* $\frac{90}{360} \times 2\pi r$ units in length.

In (iii) by the same token, minor arc AB is $\frac{\theta}{360} \times 2\pi r$ units in length.

i.e. the length of an arc of a circle, say l, is given by the following formula :—

$$l = \frac{\theta}{360} \times 2\pi r$$ where $\theta°$ is the angle subtended at the centre of the circle by the arc.

Area of a Sector

In fig. 36, AOB is called a sector of the circle. The area of the sector in (i) is $\frac{1}{2} \times \pi r^2$ i.e. $\frac{180}{360} \times \pi r^2$ units^2.

In (ii) the area of the minor sector AOB is

$\frac{1}{4} \times \pi r^2$ i.e. $\frac{90}{360} \times \pi r^2$ units^2.

In (iii) the area of minor sector AOB is

$\frac{\theta}{360} \times \pi r^2$ units^2.

Area of a Segment

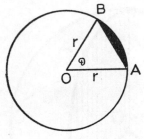

Fig. 37

In fig. 37 the shaded area is called the minor segment. The area of the minor segment is found by subtracting the area of triangle AOB from the area of the minor sector AOB i.e.
Area of segment of AOB = Area of sector AOB — Area of △ AOB.

115

Latitude and Longitude

The surface of the globe is an area, and so w
require two co-ordinates in order to fix any poi
on it. The two co-ordinates we use are longitud
and latitude.

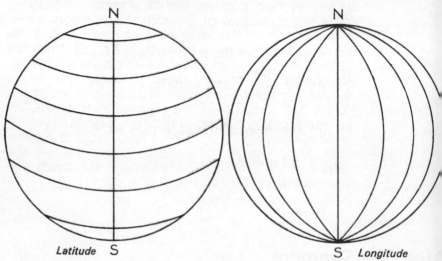

Latitude S

S *Longitude*

Longitude

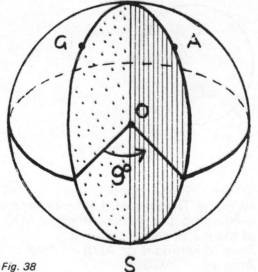

Fig. 38

Consider the globe to be a sphere with axis NOS as in fig. 38, where O is the centre of the sphere. All the semi-circular planes e.g. NOSG, NOSA have a particular longitude associated with them. The semi-circular plane which passes through Greenwich is taken as the origin, and any other plane is measured as having a rotation of g° from this position, either to the east or to the west of Greenwich. Using the point G to represent Greenwich in fig. 38, then the point A lies on the meridian of longitude g° east of Greenwich. g° is the angle between the semi-circular planes NOSG and NOSA. Any point on NOSG has longitude 0°, and any point on NOSA has longitude g° east. The range of meridians of longitude is from 0° to 180° east or west of Greenwich. The distance from O to any point on the surface of the globe is taken to be 3,960 miles, the approximate radius of the globe.

atitude

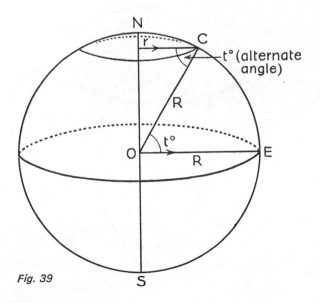

Fig. 39

Latitude is measured as the angle of elevation o depression from O, on the radius at the equator, to C, on the radius of another parallel of latitude. In fig. 39 the point C has latitude t ° north. The radius of the smaller circle of latitude is as follows:—

$r = R \cos t°$

Any point on the same circle or parallel of latitude will have the same latitude.

Latitude has a range from 0° to 90° north or south of the equator. The distance between two points on the same circle of latitude is calculated as the length of the arc of the circle between the two points by using the formula already mentioned viz

$l = \dfrac{\theta}{360} \times 2\pi r$

Here θ is the difference between the longitude o the two points, and r is the radius of the circle o latitude on which the two points lie.

Fig. 40 shows A and B on the same parallel o latitude. $\theta°$ is the difference between the longitude of A and the longitude of B.

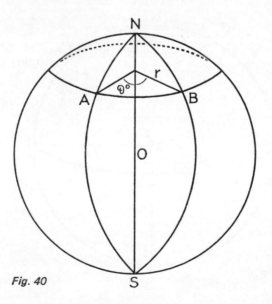

Fig. 40

Example on Latitude and Longitude

Two places A and B lie on the same circle of latitude, 56°N. A has longitude 42°E and B 21°W. Find a) the radius of the circle of latitude on which A and B lie.

b) their distance apart along the surface of the globe, on the circle of latitude.

a)
$$r = R \, \text{Cos} \, t°$$
$$= 3960 \, \text{Cos} \, 56°$$
$$= 2220 \text{ miles}$$

No	Log
3960	3·598
Cos 56 °	$\overline{1}$·748
Add	3·346

b) Length of arc

$$AB = \frac{g}{360} \times 2\pi r$$

$$g = 42 + 21$$
$$= 63$$

No	Log
63	1·799
2	0·301
3·14	0·497
2220	3·346
Add	5·943
360	2·556
subtract	3·387

Distance from A to B $= \dfrac{63}{360} \times 2 \times 3\cdot14 \times 2220$ miles.

$= 2440$ miles